# Calculation Station

*Learning Center Projects for Math*

W9-AVZ-016

Written by **Kathy Douglas**
Illustrated by **Marilynn Barr**
and **Gary Mohrman**

**Totline® Publications**
A Division of Frank Schaffer Publications, Inc.
Torrance, California

**Managing Editor:** Kathleen Cubley
**Editor:** Gayle Bittinger
**Contributing Editors:** Carol Gnojewski, Susan Hodges,
  Elizabeth McKinnon, Jean Warren
**Copyeditor:** Kris Fulsaas
**Proofreader:** Miriam Bulmer
**Editorial Assistant:** Durby Peterson
**Graphic Designer/Layout Artist:** Sarah Ness/Gordon Frazier
**Graphic Designer (Cover):** Brenda Mann Harrison
**Production Manager:** Melody Olney

ISBN: 1-57029-158-6

Library of Congress Catalog Number 97-62223
Printed in the United States of America
Published by Totline® Publications
**Editorial Office:** P.O. Box 2250
        Everett, WA 98203
**Business Office:** 23740 Hawthorne Blvd.
        Torrance, CA 90505

20 19 18 17 16 15 14 13 12 11 10 9 8 7 6 5 4 3 2

# Introduction

Working at stations is a great way to introduce your children to new concepts and reinforce ones already learned, practice skills, and have fun while learning. Station work also gives them opportunities to work independently, to remember and follow directions, and to complete projects—important skills for future learning.

The stations in this book, *Calculation Station,* are centered around developing math skills. The chapters address skills such as matching and sorting, counting, patterning, ordering, graphing, adding, and working with numbers. The final chapter has directions and patterns for a number book that each child can make.

Each station activity includes an objective for the lesson, a list of materials needed, directions for setting up the station, and step-by-step instructions for completing the project. There are also guidelines for checking the children's understanding of the concept being emphasized in the activity. The reproducible worksheets make each station quick and easy to prepare, and each one includes a skills checklist. The activities cover a wide range of abilities, so you can select the material that will best suit your children.

Most of the activities have been written as if one child were working in a station at a time. If you are planning to have more than one child working, simply adjust the materials as needed. You may also want to have an adult helper available to assist any child who may need extra guidance.

With *Calculation Station*'s easy-to-set-up stations and unique reproducible pages, you and your children will have fun with numbers all year long.

# Contents

# Matching and Sorting

# Race Car Matchups

## Objectives

*Practice matching and reading numerals.*

## Materials Needed

❏ toy cars
❏ permanent marker
❏ small box
❏ scissors
❏ pen
❏ pencil
❏ worksheets

## Setting Up the Station

- Collect ten small toy cars that you no longer need. Use a permanent marker to number the roofs of the cars from 1 to 10.
- Put the cars in a small box.
- Make five copies of the Toy Car Garage Pattern worksheet on page 9.
- Cut out the paper garages and number them from 1 to 10.
- Copy the worksheets on pages 10 and 11.
- Set out the toy cars, paper garages, a pencil, and copies of the worksheets.

## The Project

*Explain the following steps to your children.*

1. Place the paper garages on the table.
2. Select one of the cars from the box. Read the number on the roof and "drive" the car to its matching garage.
3. Repeat with the remaining cars.
4. Put the cars back in the box and pile up the paper garages.
5. Select a copy of the Matching Car worksheet and write your name on it.
6. Draw lines to connect the cars that match.
7. If there is time, write your name on a copy of the Match This worksheet and complete it by circling the two cars on the page that are exact matches.
8. Place your worksheets in the designated area.

## Follow-Up

- Call up each child and date his or her worksheets.
- Ask questions to determine the child's level of understanding.
- Further explain or demonstrate concepts as needed.
- Record the child's progress.
- Keep the worksheets in the child's folder or send them home.

# Toy Car Garage Pattern

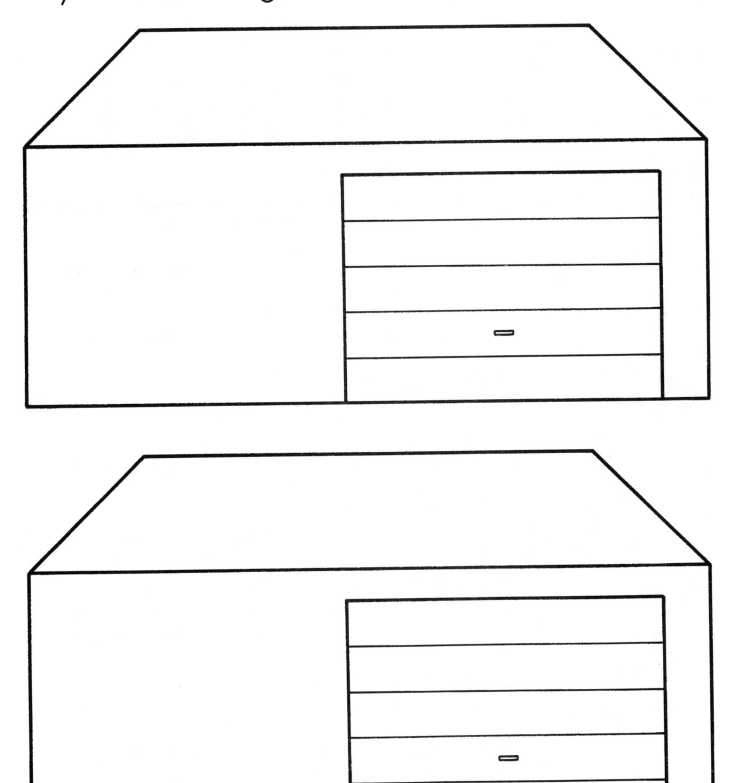

# Matching Car

Name_____

Draw lines to connect the cars that match.

Child understands: ❑ matching.

# Match This

Name_____

Color the two cars that match exactly.

Child understands: ❑ matching.

# Matching Socks

## Objectives
*Practice matching.*

## Materials Needed
❑ socks
❑ basket
❑ worksheet
❑ pencil

## Setting Up the Station
• Collect five different pairs of socks.
• Mix up the socks and put them in a basket.
• Copy the worksheet on page 13.
• Set out the basket of socks, a pencil, and copies of the worksheet.

## The Project
*Explain the following steps to your children.*

1. Pick one of the socks from the basket. Look through the other socks, find its match, and roll them together.

2. Repeat with the remaining socks.

3. Unroll the socks, put them back into the basket, and mix them up.

4. Choose a copy of the worksheet and write your name on it.

5. For each row, look at the first sock and circle the sock that matches it.

6. Place your worksheet in the designated area.

## Follow-Up
• Call up each child and date his or her worksheet.
• Ask questions to determine the child's level of understanding.
• Further explain or demonstrate concepts as needed.
• Record the child's progress.
• Keep the worksheet in the child's folder or send it home.

# Find the Match

Name_____

In each row, circle the sock that matches the first one.

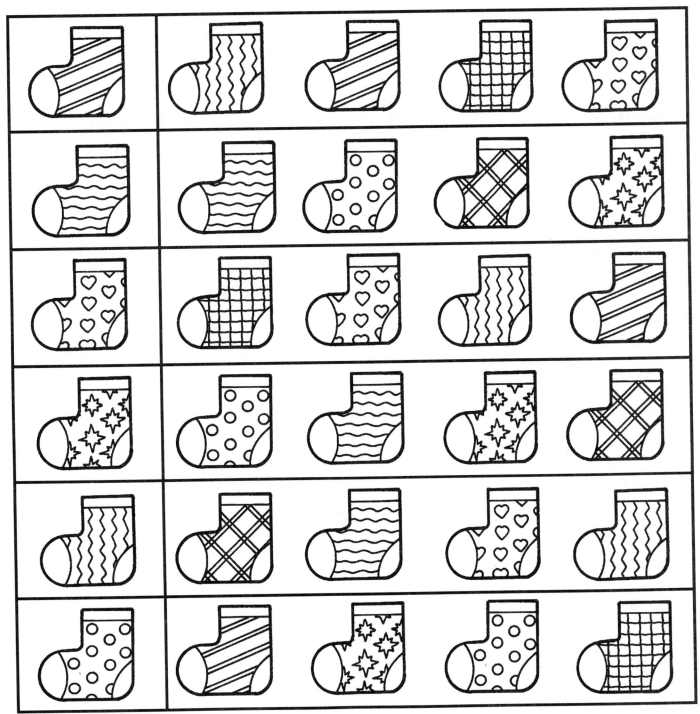

Child understands: ❏ matching.

*Matching and Sorting* • Calculation Station

# Big and Little

## Objectives

*Practice sorting and counting, and develop an understanding of big and little.*

## Materials Needed

❑ big and little objects
❑ box
❑ construction paper
❑ pen
❑ pencil
❑ crayons
❑ worksheet

## Setting Up the Station

- Collect big and little objects that are similar. For example, you could find a big spoon and a little spoon, a big pencil and a little pencil, and a big doll and a small doll.
- Put the objects in a box.
- Write the word "big" on a sheet of construction paper and the word "little" on another sheet. If you wish, write the word "big" in very big letters and the word "little" in very tiny letters.
- Copy the worksheet on page 15.
- Set out the box of big and little objects, the pages with "big" and "little" written on them, a pencil, crayons, and copies of the worksheet.

## The Project

*Explain the following steps to your children.*

1. Set out the pages with "big" and "little" written on them.
2. Take the objects out of the box and sort them by placing the big objects on the "big" page and the little objects on the "little" page.
3. Put the objects back in the box.
4. Take a copy of the worksheet and write your name on it.
5. On the top half of the page, color the object that is big. On the bottom half of the page, color the object that is little.
6. Place your worksheet in the designated area.

## Follow-Up

- Call up each child and date his or her worksheet.
- Ask questions to determine the child's level of understanding.
- Further explain or demonstrate concepts as needed.
- Record the child's progress.
- Keep the worksheet in the child's folder or send it home.

# Big or Little

Name_____

Color the one that is big.

Color the one that is little.

Child understands: ❏ big and little.

# Star Sort

## Objectives

*Practice sorting and counting, and develop an understanding of small, medium, and large.*

## Materials Needed

❑ yellow construction paper
❑ scissors
❑ box
❑ pencil
❑ crayons
❑ worksheets

## Setting Up the Station

• Using the Star Patterns worksheet on page 17 as a guide, cut small, medium, and large star shapes out of yellow construction paper, at least three of each size.

• Place the stars in a box.

• Copy the worksheets on pages 18 and 19.

• Set out the stars, a pencil, crayons, and copies of the worksheets.

## The Project

*Explain the following steps to your children.*

1. Take the stars out of the box and sort them by size into three groups.

2. Count the number of stars in each group.

3. Mix up the stars and put them back in the box.

4. Select a copy of the Stars in the Sky worksheet and write your name on it.

5. Color all of the big stars on the page.

6. Select a copy of the Small, Medium, and Large worksheet and write your name on it.

7. Follow the directions for each line and circle the stars of the appropriate size.

8. Place your worksheets in the designated area.

## Follow-Up

• Call up each child and date his or her worksheets.

• Ask questions to determine the child's level of understanding.

• Further explain or demonstrate concepts as needed.

• Record the child's progress.

• Keep the worksheets in the child's folder or send them home.

# Star Patterns

# Stars in the Sky

Name_____

Color the big stars.

Child understands: ❑ sorting; ❑ big and little.

# Small, Medium, and Large

Name_____

Circle the small ⭐s.

Circle the medium ⭐s.

Circle the large ⭐s.

Child understands: ❏ sorting; ❏ small, medium, and large.

# Quilt Squares

## Objectives

*Practice sorting and counting.*

## Materials Needed

❏ fabric
❏ ruler
❏ scissors
❏ basket
❏ pencil
❏ crayons
❏ worksheet

## Setting Up the Station

• Select fabric in four different patterns. Cut several 5-inch squares out of each kind of fabric to make "quilt" squares. (You do not have to have the same number of squares from each fabric pattern.)

• Mix up the quilt squares and place them in a basket.

• Copy the worksheet on page 21.

• Set out the basket of quilt squares, a pencil, crayons, and copies of the worksheet.

## The Project

*Explain the following steps to your children.*

1. Take the quilt squares out of the basket and look at them.

2. Sort the squares by patterns into four different groups.

3. Count the number of quilt squares in each group.

4. Mix up the quilt squares and return them to the basket.

5. Take a copy of the worksheet and write your name on it.

6. Following the color key on the worksheet, color the large quilt.

7. Place your completed worksheet in the designated area.

## Follow-Up

• Call up each child and date his or her worksheet.

• Ask questions to determine the child's level of understanding.

• Further explain or demonstrate concepts as needed.

• Record the child's progress.

• Keep the worksheet in the child's folder or send it home.

# Colorful Quilt

Color the △ squares red.
Color the ☆ squares blue.
Color the ✿ squares yellow.

Child understands: ❑ sorting; ❑ matching.

# Counting

# Flower Power

## Objectives

*Practice counting and reading numerals.*

## Materials Needed

❑ permanent marker
❑ paper cups
❑ modeling dough
❑ paper or silk flowers
❑ box
❑ crayons
❑ worksheets

## Setting Up the Station

- Use a permanent marker to number five paper cups from 1 to 5 *or* from 6 to 10. Put a small amount of modeling dough in the bottom of each of the cups to prevent them from tipping over when filled with flowers.
- Set out the cups in sequential order.
- Place 15 paper or silk flowers in a box if you are working with the numbers 1 through 5; put 40 flowers in the box if you are working with the numbers 6 through 10.
- Copy the worksheets on pages 25–27.
- Set out the paper cups, box of flowers, crayons, and copies of the worksheets.

## The Project

*Explain the following steps to your children.*

1. Identify the numbers on the paper cups and place the correct number of flowers in each cup.

2. Put the flowers back in the box.

3. Select a copy of one of the Flowers worksheets and write your name on it.

4. On the worksheet, identify the number written on each flower pot and draw the appropriate number of flowers in each one.

5. If there is time, write your name on a copy of the Counting Flowers worksheet and complete it by counting the flowers in each pot and drawing a line to the correct numeral.

6. Place your worksheets in the designated area.

## Follow-Up

- Call up each child and date his or her worksheets.
- Ask questions to determine the child's level of understanding.
- Further explain or demonstrate concepts as needed.
- Record the child's progress.
- Keep the worksheets in the child's folder or send them home.

# Flowers: 1 to 5

Name_____

Draw the correct number of flowers in each flower pot.

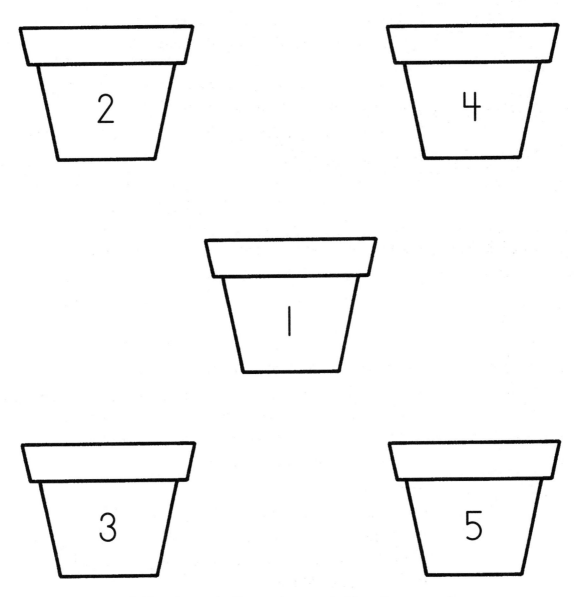

Child understands: ❏ counting 1 to 5; ❏ reading numerals.

# Flowers: 6 to 10

Name_____

Draw the correct number of flowers in each flower pot.

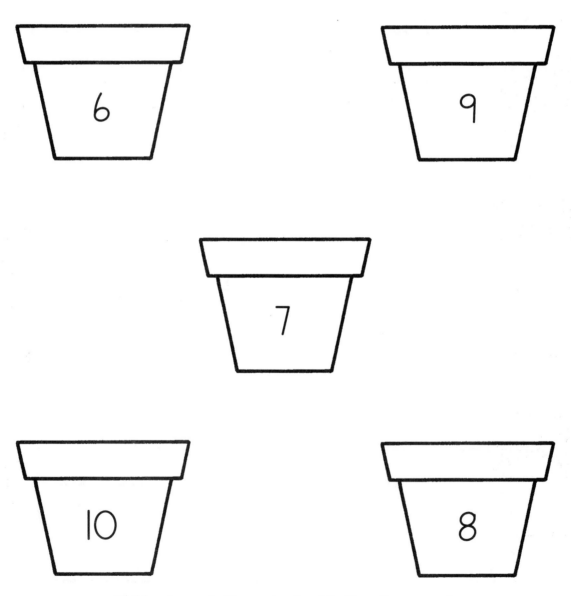

Child understands: ❑ counting 6 to 10; ❑ reading numerals.

# Counting Flowers

Name_____

Count the flowers in each pot.
Draw a line to the correct numeral.

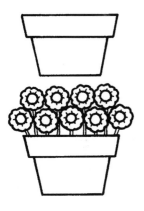

0

9

3

5

7

Child understands: ❑ counting 1 to 10; ❑ reading numerals.

# Feather Count

## Objectives

*Practice counting, sorting, and reading and writing numerals.*

## Materials Needed

❑ craft feathers
❑ box
❑ flat, rubber eraser
❑ sharp knife
❑ wooden block (optional)
❑ glue (optional)
❑ ink pad
❑ pencil
❑ worksheets

## Setting Up the Station

• Collect craft feathers in a variety of colors. Put them in a box.

• Make a feather stamp by using a sharp knife to cut a flat, rubber pencil eraser into a feather shape. If you wish, glue the rubber feather shape onto a wooden block to make it easier to handle. (Or, use a readymade feather stamp.)

• Copy the worksheets on pages 29–31.

• Set out the feather stamp, an ink pad, a pencil, and copies of the worksheets.

## The Project

*Explain the following steps to your children.*

1. Take a handful of feathers out of the box.

2. Sort the feathers by color, putting each color in a separate pile.

3. Count how many feathers of each color you have.

4. Put the feathers back in the box.

5. Select a copy of one of the Turkeys worksheets and write your name on it.

6. Look at the number on one of the turkeys and use the feather stamp to print that many feathers on the turkey for a tail. Repeat for the remaining turkeys.

7. If there is time, write your name on a copy of the Counting Feathers worksheet and complete it by counting the feathers on each turkey and writing the appropriate numeral on the line.

8. Place your worksheets in the designated area.

## Follow-Up

• Call up each child and date his or her worksheets.

• Ask questions to determine the child's level of understanding.

• Further explain or demonstrate concepts as needed.

• Record the child's progress.

• Keep the worksheets in the child's folder or send them home.

# Turkeys: 1 to 5

Name_____

Stamp the correct number of feathers on the turkey for a tail.

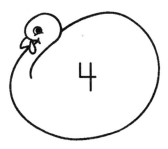

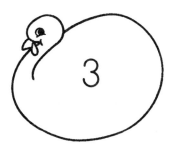

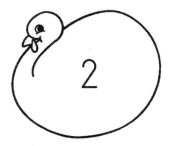

Child understands: ❑ counting 1 to 5; ❑ reading numerals.

# Turkeys: 6 to 10

Name_____

Stamp the correct number of feathers on the turkey for a tail.

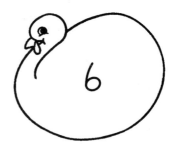

Child understands: ❏ counting 6 to 10; ❏ reading numerals.

# Counting Feathers

Name_____

Count the feathers on each turkey. Write the numeral on the line.

Child understands: ❑ counting 1 to 10; ❑ writing numerals.

# Colored Cubes

## Objectives

*Practice counting and reading numerals.*

## Materials Needed

❑ cube manipulatives
❑ construction paper
❑ ruler
❑ scissors
❑ felt tip marker
❑ crayons
❑ pencil
❑ worksheets

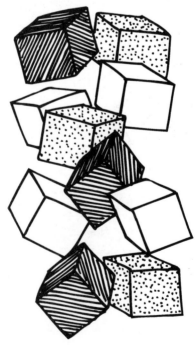

## Setting Up the Station

• Collect a tub or box of colored cube manipulatives.
• Cut ten 4-inch squares out of construction paper and use a felt tip marker to number the squares from 1 to 10.
• Copy the worksheets on pages 33–35.
• Set out the cube manipulatives, construction paper squares, crayons, a pencil, and copies of the worksheets.

## The Project

*Explain the following steps to your children.*

1. Select one of the construction paper squares. Identify the number on the paper and count out that many colored cubes. Set the square aside and choose another square. Repeat for all of the squares.

2. Put the cubes back in the tub and stack the paper squares together.

3. Select a copy of one of the Cube Count worksheets and write your name on it.

4. Look at the number at the beginning of each row. Use the crayons to color in that number of cubes.

5. If there is time, write your name on a copy of the Cube Building worksheet. Count the cubes in each building and circle the correct number.

6. Place your worksheets in the designated area.

## Follow-Up

• Call up each child and date his or her worksheets.
• Ask questions to determine the child's level of understanding.
• Further explain or demonstrate concepts as needed.
• Record the child's progress.
• Keep the worksheets in the child's folder or send them home.

# Cube Count: 1 to 5

Name_____

For each row, say the number. Then color that many cubes.

1

4

2

5

3

Child understands: ❑ counting 1 to 5; ❑ reading numerals.

# Cube Count: 6 to 10

Name_____

For each row, say the number. Then color that many cubes.

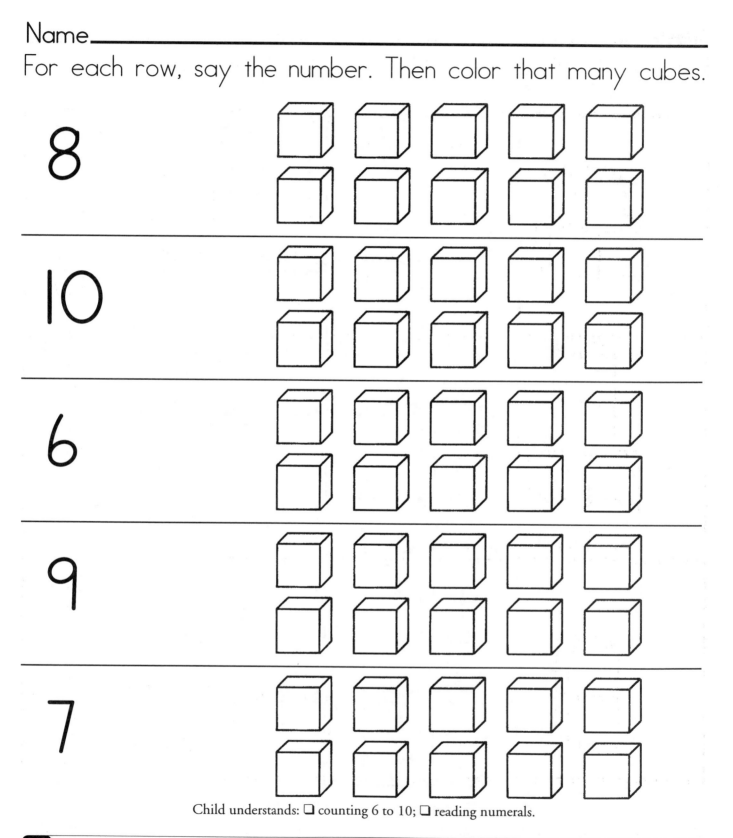

8

10

6

9

7

Child understands: ❑ counting 6 to 10; ❑ reading numerals.

# Cube Building

Name_____

For each building, count the cubes. Circle the correct number.

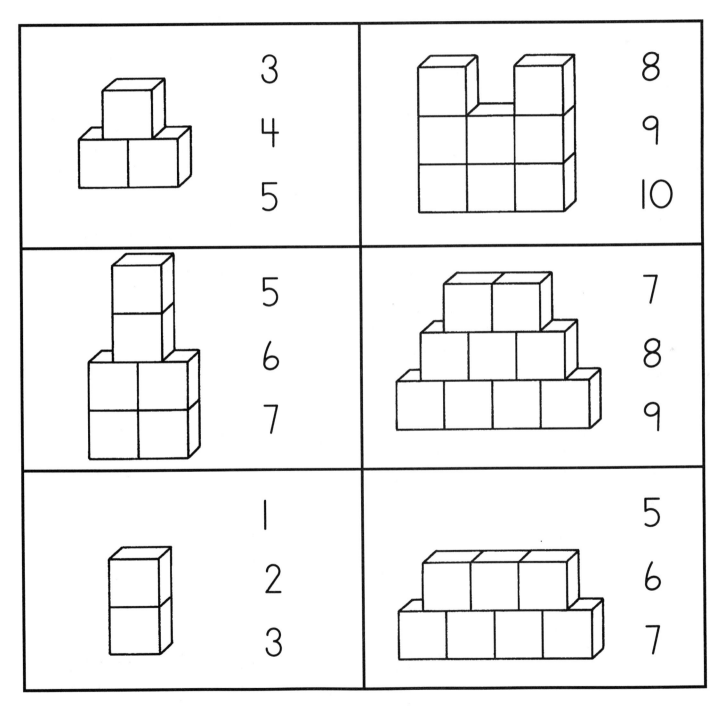

3
4
5

8
9
10

5
6
7

7
8
9

1
2
3

5
6
7

Child understands: ❏ counting 1 to 10; ❏ reading numerals.

# Number Necklace

## Objectives

*Practice reading and writing numerals and counting.*

## Materials Needed

❑ O-shaped cereal
❑ bowl
❑ construction paper
❑ ruler
❑ scissors
❑ hole punch
❑ pen
❑ yarn
❑ tape
❑ pencil
❑ worksheet

## Setting Up the Station

- Purchase a box of fruit-flavored, O-shaped cereal. Pour some of the cereal into a bowl.
- Cut 3-inch circles out of construction paper. Use a hole punch to make a hole at the top of each circle.
- Decide on a number and write it on each paper circle.
- Cut an 18-inch length of yarn for each child. Thread one of the paper circles onto each yarn length and tape the other end of the yarn to make a "needle."
- Copy the worksheet on page 37.
- Set out the bowl of cereal, lengths of yarn, a pencil, and copies of the worksheet.

## The Project

*Explain the following steps to your children.*

1. Select one of the lengths of yarn and write your name on the circle.
2. Look at the number on the circle and count out that many pieces of fruit-flavored cereal.
3. String the cereal on your length of yarn.
4. Tie the ends together to make a necklace.
5. Select a copy of the Counting Beads worksheet and write your name on it.
6. Count the number of beads on each necklace and write the number on the line.
7. Place your worksheet in the designated area.

## Follow-Up

- Call up each child and date his or her worksheet.
- Ask questions to determine the child's level of understanding.
- Further explain or demonstrate concepts as needed.
- Record the child's progress.
- Keep the worksheet in the child's folder or send it home.

# Counting Beads

Name_____

Count the beads on each necklace. Write that number on the line.

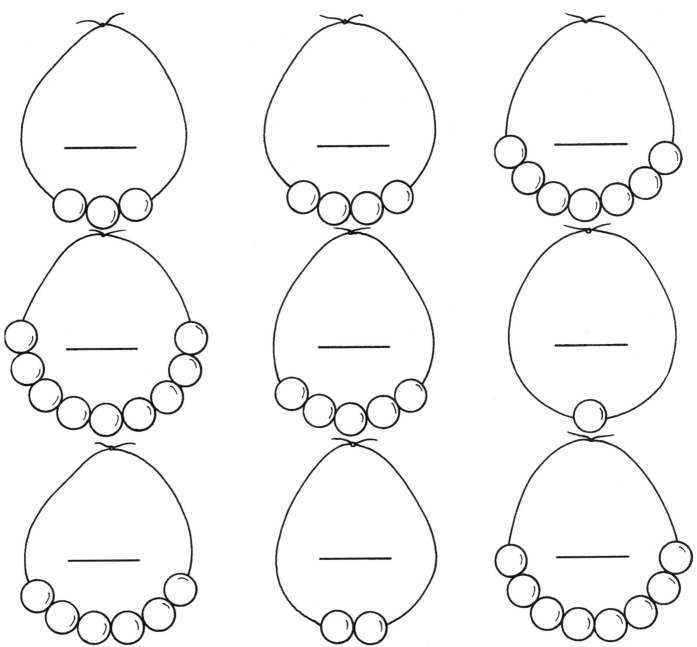

Child understands: ❑ counting 1 to 10; ❑ writing numerals.

# Fun Stamps

## Objectives

*Practice counting, and reading and writing numerals.*

## Materials Needed

❑ rubber stamps
❑ ink pad
❑ pencil
❑ worksheets

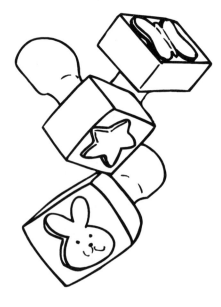

## Setting Up the Station

- Collect small rubber stamps of various designs.
- Copy the worksheets on pages 39–41.
- Set out the rubber stamps, an ink pad, a pencil, and copies of the worksheets.

## The Project

*Explain the following steps to your children.*

1. Choose one of the rubber stamps.
2. Select a copy of one of the Counting Stamps worksheets and write your name on it.
3. Read the numeral in each circle.
4. Use the rubber stamp to make a print that many times in the circle.
5. If there is time, write your name on a copy of the Stamp Your Own Number worksheet and complete it by making rubber stamp prints, counting them, and writing the numeral in the box.
6. Place your worksheets in the designated area.

## Follow-Up

- Call up each child and date his or her worksheets.
- Ask questions to determine the child's level of understanding.
- Further explain or demonstrate concepts as needed.
- Record the child's progress.
- Keep the worksheets in the child's folder or send them home.

# Counting Stamps: 1 to 5

Name_____

Read the number in each circle. Make that many rubber stamp prints in the circle.

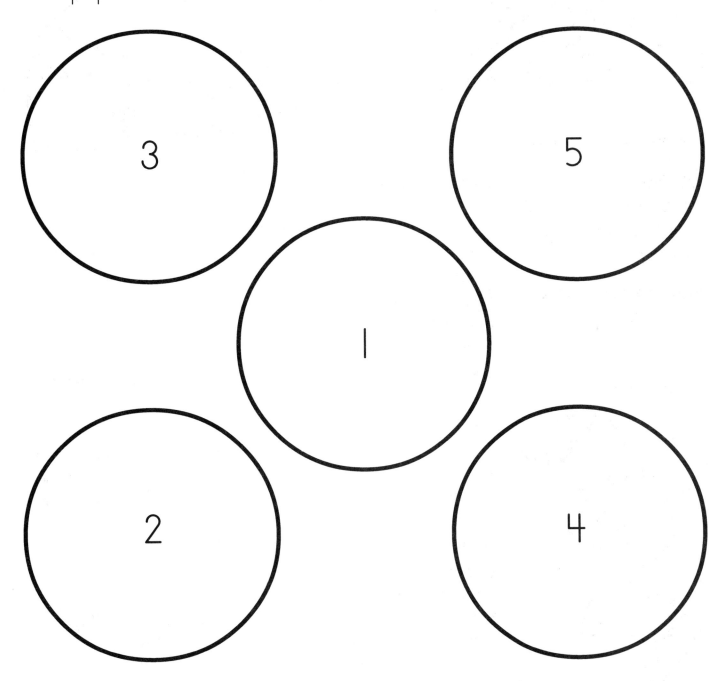

Child understands: ❏ counting 1 to 5; ❏ reading numerals.

# Counting Stamps: 6 to 10

Name_____

Read the number in each circle. Make that many rubber stamp prints in the circle.

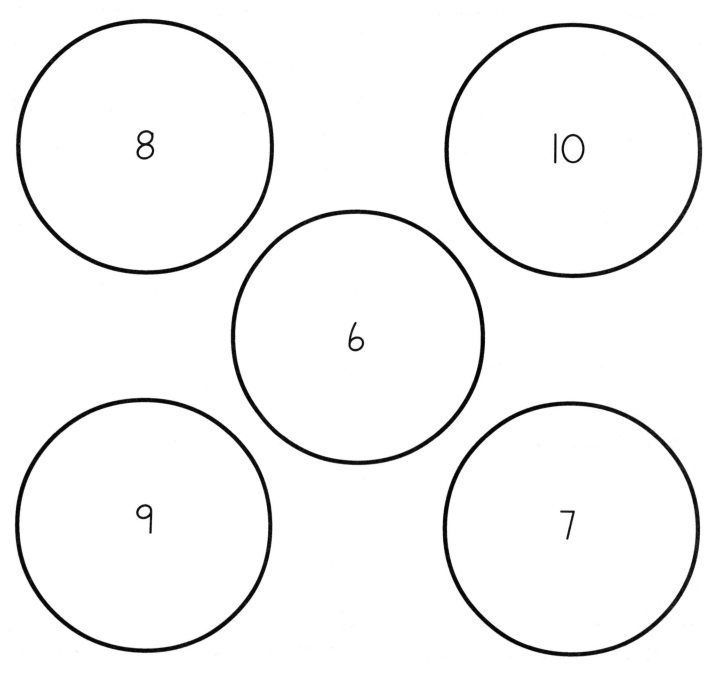

Child understands: ❏ counting 6 to 10; ❏ reading numerals.

# Stamp Your Own Number

Name_____

Make rubber stamp prints. Count the prints. Write that number in the box.

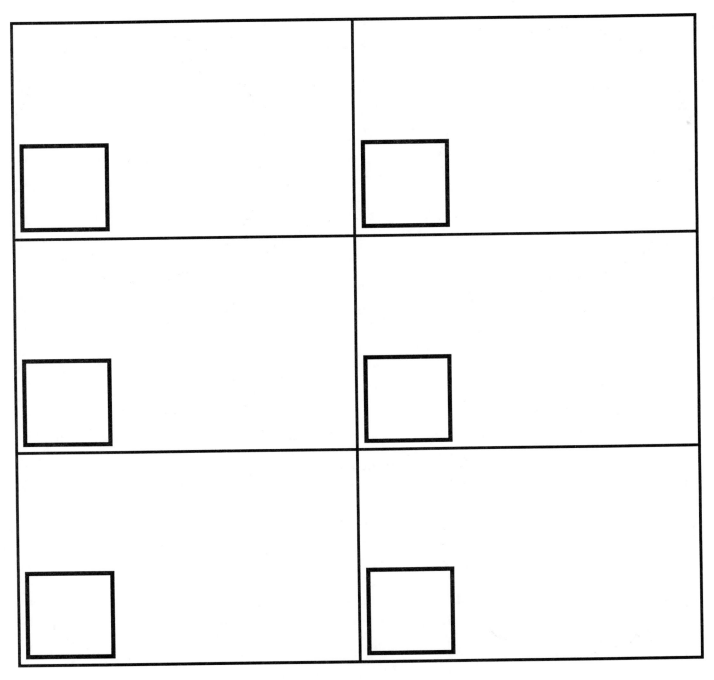

Child understands: ❑ counting; ❑ writing numerals.

# Cupcake Birthday Fun

## Objectives

*Practice counting, reading numerals, and matching.*

## Materials Needed

❑ index cards
❑ pen
❑ modeling dough
❑ birthday candles
❑ scissors
❑ pencil
❑ crayons
❑ worksheets

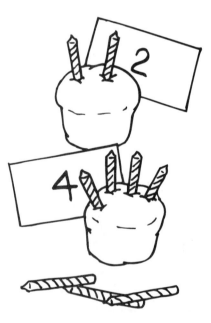

## Setting Up the Station

• Count out ten index cards. Number the cards from 1 to 10.
• Prepare a batch of modeling dough or purchase readymade dough.
• Cut the wicks off ten birthday candles.
• Copy the worksheets on pages 43-45.
• Set out the index cards, modeling dough, birthday candles, a pencil, crayons, and copies of the worksheets.

## The Project

*Explain the following steps to your children.*

1. Form a cupcake out of the modeling dough.
2. Choose one of the index cards and read the number on it.
3. Count out that many birthday candles.
4. Put the candles in your cupcake.
5. Take out the candles and put the dough away.
6. Select a copy of one of the Cupcake Candles worksheets and write your name on it.
7. Read the number written on the cupcake and draw on that many candles.
8. If there is time, write your name on a copy of the Matching Cupcakes worksheet and complete it by drawing lines between the cupcakes that have the same number of candles.
9. Place your worksheets in the designated area.

## Follow-Up

• Call up each child and date his or her worksheets.
• Ask questions to determine the child's level of understanding.
• Further explain or demonstrate concepts as needed.
• Record the child's progress.
• Keep the worksheets in the child's folder or send them home.

# Cupcake Candles: 1 to 5

Name_____

Read the number on each cupcake. Draw that many candles.

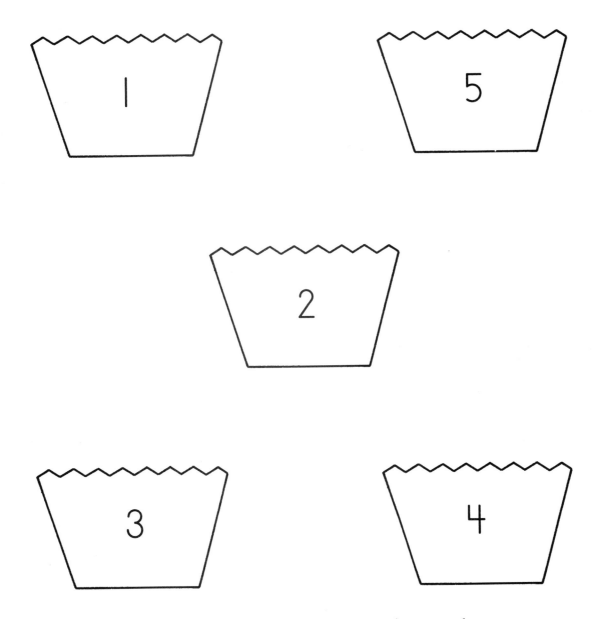

Child understands: ❑ counting 1 to 5; ❑ reading numerals.

# Cupcake Candles: 6 to 10

Name_____

Read the number on each cupcake. Draw that many candles.

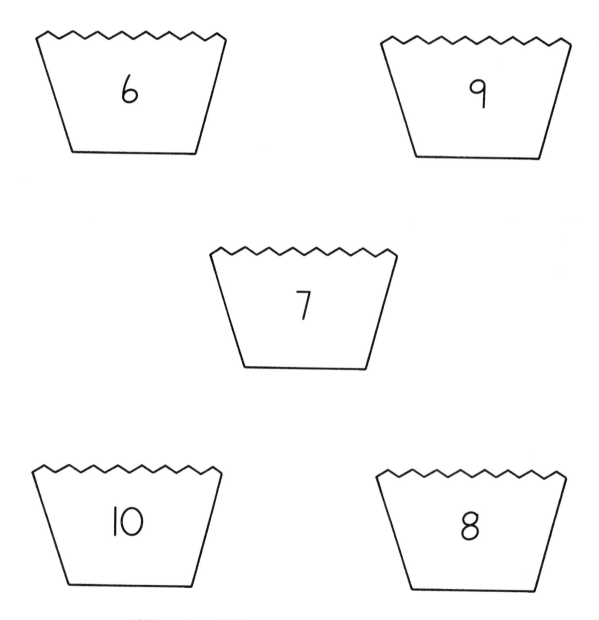

Child understands: ❑ counting 6 to 10; ❑ reading numerals.

# Matching Cupcakes

Name_____

Draw a line between the cupcakes with the same number of candles.

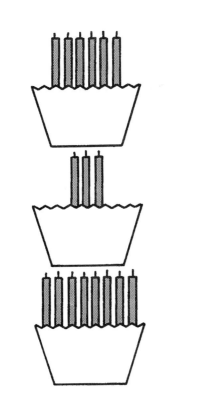
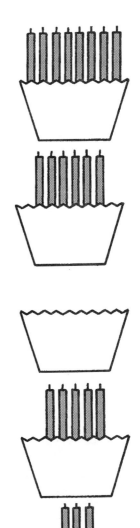
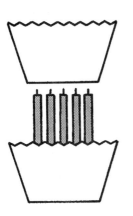

Child understands: ❑ counting; ❑ matching.

# Which Treasure?

## Objectives

*Practice sorting and counting, and understand the concept of more and less.*

## Materials Needed

❑ small boxes
❑ shiny gift-wrap
❑ tape
❑ scissors
❑ small items
❑ pencil
❑ worksheet

## Setting Up the Station

- Cover four small cardboard jewelry boxes and lids with shiny gift-wrap to make treasure boxes. (Be sure the lids are wrapped separately so they can be taken on and off.)
- Collect two different kinds of small items such as pennies and pebbles, or beads and pompoms.
- Put a different amount of each item into each box, so that in each box there is more of one "treasure" than the other.
- Copy the worksheet on page 47.
- Set out the treasure boxes, a pencil, and copies of the worksheet.

## The Project

*Explain the following steps to your children.*

1. Choose one of the treasure chests.

2. Empty out the treasure and sort it into two piles.

3. Count how many of each treasure you have. Which treasure is more? Which treasure is less?

4. Repeat with the remaining treasure boxes.

5. Take a copy of the worksheet and write your name on it.

6. Look at the treasures in each box . Circle the treasure that is more. Put an X over the treasure that is less.

7. Place your worksheet in the designated area.

## Follow-Up

- Call up each child and date his or her worksheet.
- Ask questions to determine the child's level of understanding.
- Further explain or demonstrate concepts as needed.
- Record the child's progress.
- Keep the worksheet in the child's folder or send it home.

# More or Less?

Name_____

In each box, circle the treasure that is more. Put an X over the treasure that is less.

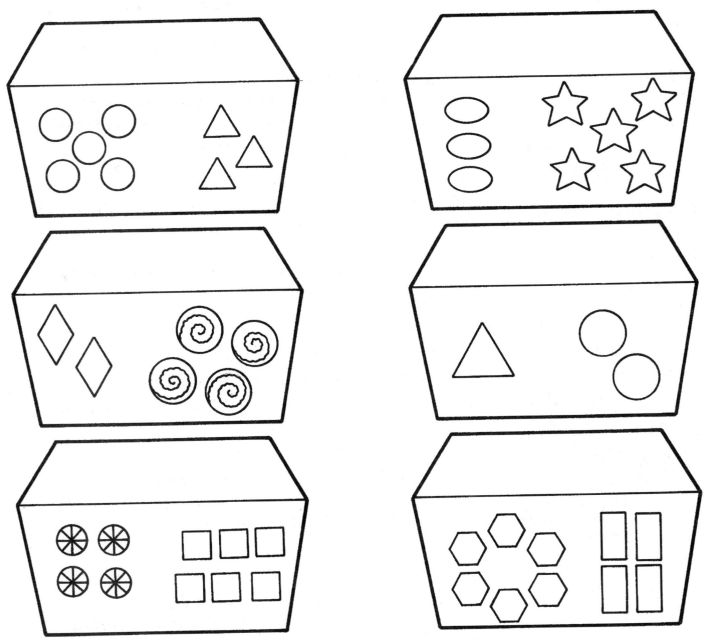

Child understands: ❑ counting; ❑ concept of more and less.

# Patterning

# Shape Patterns

## Objectives

*Practice recognizing, copying, and making patterns, and reinforce the concept of shapes.*

## Materials Needed

❑ worksheets
❑ construction paper
❑ scissors
❑ paper strips
❑ pen
❑ pencil

## Setting Up the Station

- Using the Shape Patterns worksheet on page 51 as a guide, cut triangles, circles, and squares out of construction paper, at least ten of each shape.
- On a long strip of paper, arrange the shapes in a two-part pattern (circle-square, circle-square or triangle-circle, triangle-circle). Trace around each shape before removing it from the paper. Make two or three different pattern strips. You may also wish to make pattern strips with three-part patterns (triangle-square-square, triangle-square-square).
- Copy the worksheets on pages 52–53.
- Set out the paper shapes, pattern strips, a pencil, and copies of the worksheets.

## The Project

*Explain the following steps to your children.*

1. Select one of the pattern strips and look at the pattern.
2. Find the paper shapes shown on the pattern. Arrange them on top of the pattern strip to make the pattern. Then use more paper shapes to continue the pattern.
3. If you wish, use the paper shapes to make your own patterns.
4. Choose a copy of one of the What's Next? worksheets and write your name on it.
5. Look at each row, identify the pattern, and draw the shape that comes next.
6. Place your worksheet in the designated area.

## Follow-Up

- Call up each child and date his or her worksheet.
- Ask questions to determine the child's level of understanding.
- Further explain or demonstrate concepts as needed.
- Record the child's progress.
- Keep the worksheet in the child's folder or send them home.

# Shape Patterns

# What's Next? Two-Part Patterns

Name_____

For each row, draw the shape that comes next.

Child understands: ☐ two-part patterns; ☐ shapes.

# What's Next? Three-Part Patterns

Name_____

For each row, draw the shape that comes next.

Child understands: ❑ three-part patterns; ❑ shapes.

# Teddy Bear Patterns

## Objectives

*Practice recognizing, copying, and making patterns.*

## Materials Needed

❑ worksheets
❑ scissors
❑ box
❑ pencil
❑ glue

## Setting Up the Station

- Make several copies of the Pattern Bears worksheet on page 55. Use scissors or a paper cutter to cut out the individual bear pattern squares.
- Place the squares, all mixed up, in a box.
- Copy the worksheets on pages 56–57.
- Set out the box of bear pattern squares, a pencil, glue, and copies of the worksheets.

## The Project

*Explain the following steps to your children.*

1. Take a few of the bear pattern squares out of the box.

2. Use the squares to make a simple pattern, such as sitting-standing, sitting-standing.

3. Select a copy of the Fill in the Box worksheet and write your name on it.

4. Look at the pattern in each row and decide which bear comes next.

5. Find that bear pattern square and glue it in the dotted box on the worksheet.

6. If you have time, take a copy of the Make a Pattern worksheet and write your name on it. Complete the worksheet by finding the bear pattern squares that match the patterns, and gluing them in order on the worksheet.

7. Place your worksheets in the designated area.

## Follow-Up

- Call up each child and date his or her worksheets.
- Ask questions to determine the child's level of understanding.
- Further explain or demonstrate concepts as needed.
- Record the child's progress.
- Keep the worksheets in the child's folder or send them home.

# Pattern Bears

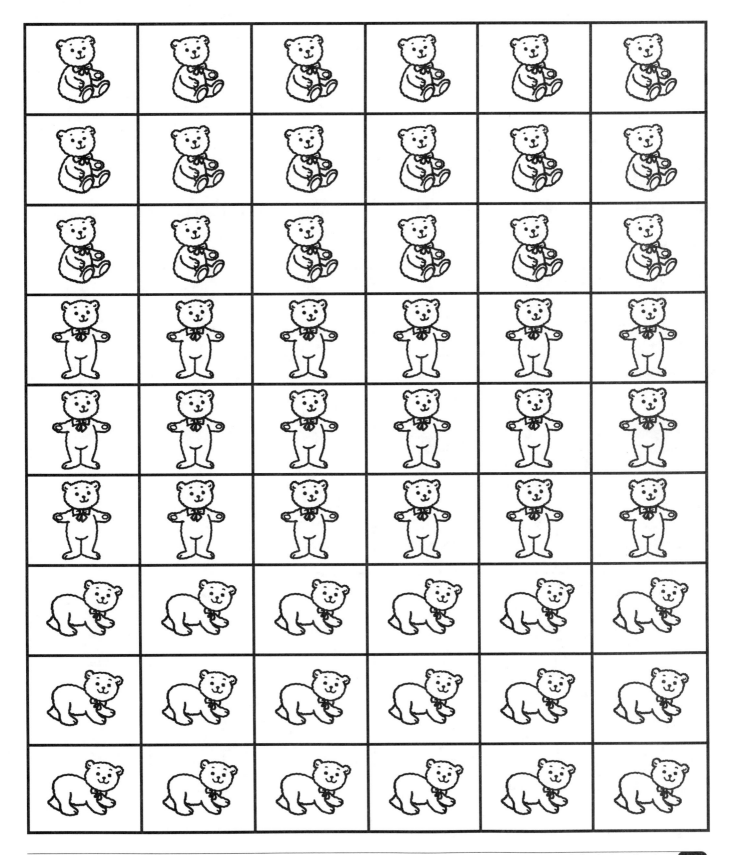

# Fill in the Box

Name_____

For each row, decide what bear comes next. Glue that
bear pattern square in the dotted box.

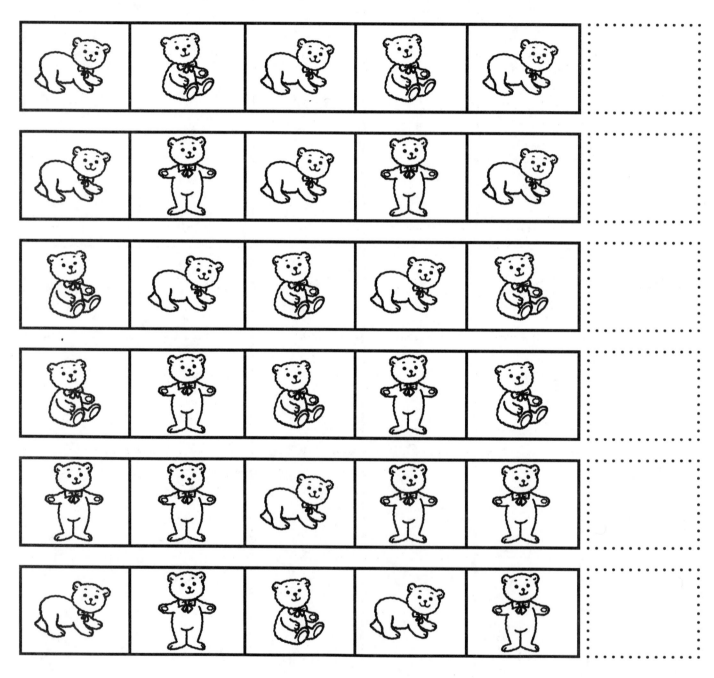

Child understands: ❑ continuing patterns.

# Make a Pattern

Name_____

Glue bear pattern squares in the dotted boxes to copy each pattern.

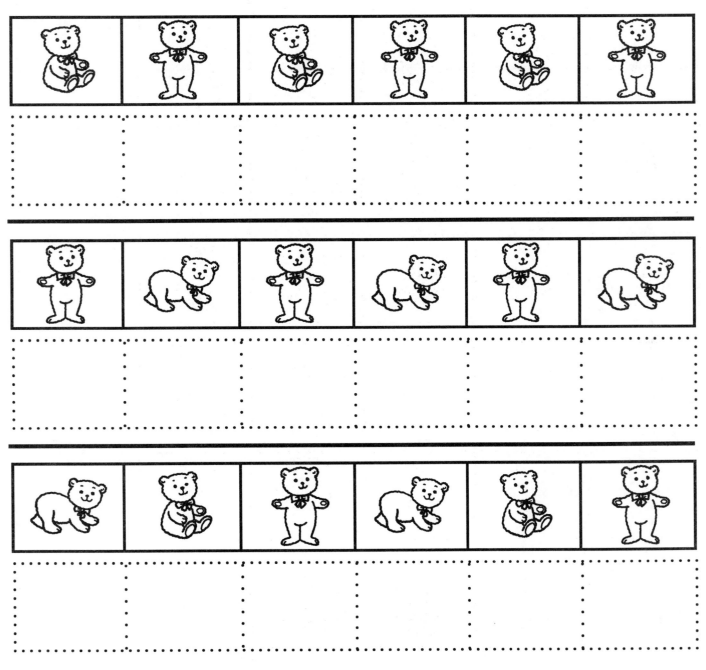

Child understands: ❑ copying patterns.

# Button Patterns

## Objectives
*Practice identifying, continuing, and making patterns.*

## Materials Needed
❑ buttons
❑ box
❑ construction paper
❑ scissors
❑ glue
❑ pencil
❑ crayons
❑ worksheets

## Setting Up the Station
• Collect buttons of different colors and sizes, and with different numbers of holes.
• Put the buttons in a box.
• Cut a piece of construction paper lengthwise into three strips.
• On one of the paper strips, glue several buttons in a color pattern. On another paper strip, glue buttons in a size pattern. For the third paper strip, find a few buttons with four holes and a few with two holes, then glue the buttons on the paper to make a pattern.
• Display the sample patterns in the station.
• Copy the worksheets on pages 59–61.
• Set out the box of buttons, a pencil, crayons, and copies of the worksheets.

## The Project
*Explain the following steps to your children.*

1. Look at the button patterns in the station.

2. Examine the buttons in the box.

3. Use the buttons to make a color pattern, a size pattern, and a number-of-holes pattern.

4. Select a copy of the Color Patterns worksheet and write your name on it. Complete the worksheet by coloring the buttons to make a pattern.

5. Choose a copy of the Size Patterns worksheet and write your name on it. Complete the worksheet by drawing the next button in each pattern.

6. Take a copy of the Number Patterns worksheet and write your name on it. For each row, count the holes in the buttons. Determine the pattern and add the correct number of holes to the last button to complete the pattern.

7. Place your worksheets in the designated area.

## Follow-Up
• Call up each child and date his or her worksheets.
• Ask questions to determine the child's level of understanding.
• Further explain or demonstrate concepts as needed.
• Record the child's progress.
• Keep the worksheets in the child's folder or send them home.

# Color Patterns

Name_____

Color the buttons to make a pattern.

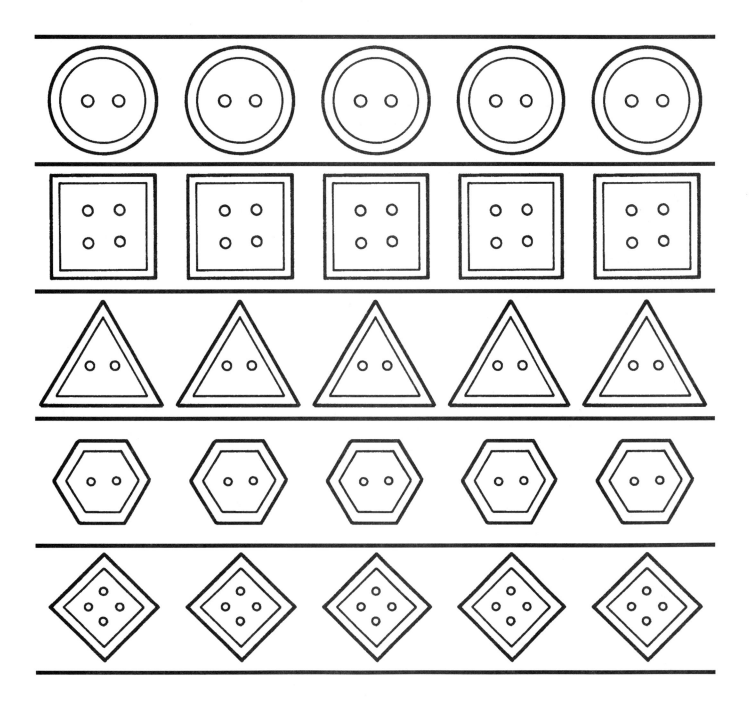

Child understands: ❑ making patterns.

# Size Patterns

Name_____

Draw the size of button that comes next in each pattern.

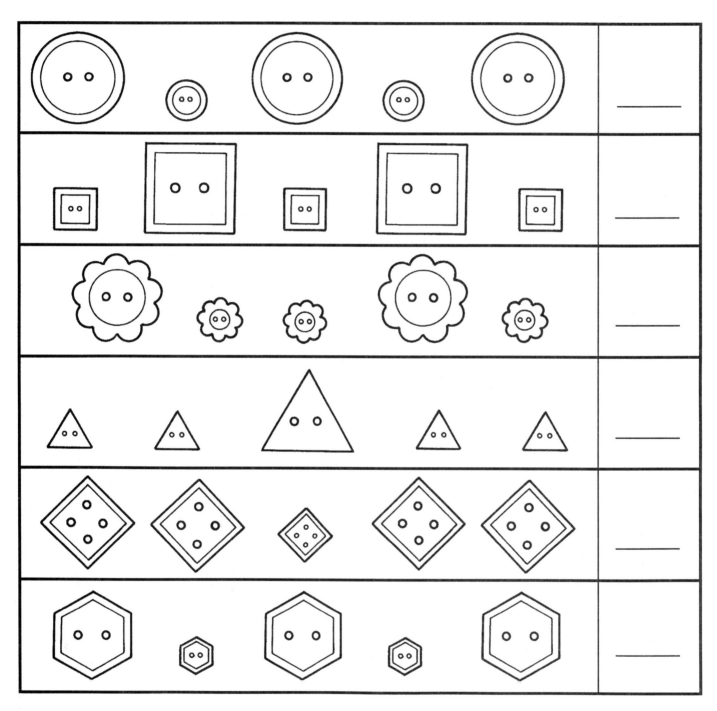

Child understands: ❏ continuing patterns.

# Number Patterns

Name_____

For each row, count the holes in the buttons to figure out the pattern. Draw the holes on the last button to complete the pattern.

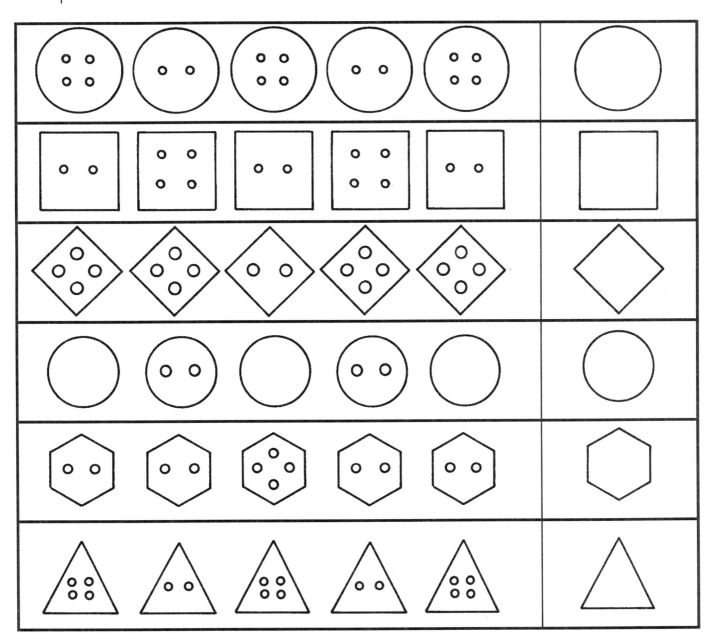

Child understands: ❑ continuing patterns.

# Ordering

# Gifts of All Sizes

## Objectives

*Practice putting objects in order of size.*

## Materials Needed

☐ four boxes
☐ gift-wrap
☐ bows (optional)
☐ tape
☐ scissors
☐ glue
☐ pencil
☐ worksheet

## Setting Up the Station

• Collect four boxes of various sizes.
• Wrap the boxes with gift-wrap. Add bows, if you wish.
• Arrange the boxes in the station in a random order.
• Copy the worksheet on page 65.
• Set out the glue, scissors, a pencil, and copies of the worksheet.

## The Project

*Explain the following steps to your children.*

1. Arrange the gift boxes in a line from the smallest to the largest.

2. Mix up the boxes and set them aside.

3. Select a copy of the worksheet and write your name on it.

4. Cut off the bottom of the worksheet on the dotted line, then cut out the individual presents.

5. In the blank squares on the worksheet, arrange the presents from the smallest to the largest, and glue them in place.

6. Place your worksheet in the designated area.

## Follow-Up

• Call up each child and date his or her worksheet.
• Ask questions to determine the child's level of understanding.
• Further explain or demonstrate concepts as needed.
• Record the child's progress.
• Keep the worksheet in the child's folder or send it home.

# Sizing Up Gifts

Name_____

Cut out the presents and glue them in the boxes in order from smallest to largest.

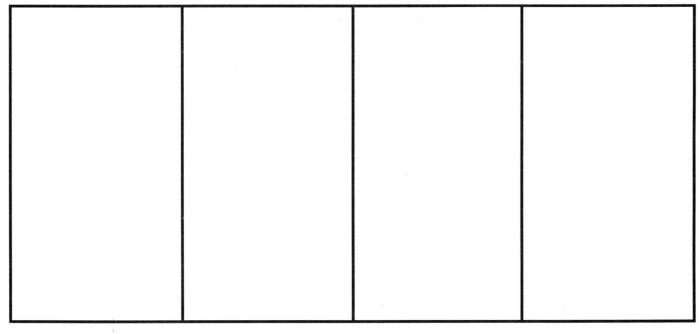

Child understands: ❑ ordering by size.

✂ - - - - - - - - - - - - - - - - - - - - - - - - - - - - - - - - - - - - - - - -

# Here's the Story

## Objectives

*Practice arranging stories in chronological order.*

## Materials Needed

❏ construction paper
❏ ruler
❏ scissors
❏ glue
❏ pencil
❏ worksheets

## Setting Up the Station

• Cut 6-by-18-inch strips out of construction paper.
• Copy the worksheets on pages 67–69.
• Set out the paper strips, scissors, glue, a pencil, and copies of the worksheets.

## The Project

*Explain the following steps to your children.*

1. Select a copy of one of the worksheets.

2. Cut out the four story squares.

3. Arrange the squares in order on one of the paper strips.

4. Glue the squares in place. Write your name on the back of your "story strip."

5. If time allows, do the other worksheets in the same way.

6. Place your completed story strips in the designated area.

## Follow-Up

• Call up each child and date his or her story strips.
• Ask questions to determine the child's level of understanding.
• Further explain or demonstrate concepts as needed.
• Keep the story strips in the child's folder or send them home.

# Baking Cookies

Name_____

Cut out the squares. Glue the squares in order on a paper strip.

# Drawing a Picture

Name_____

Cut out the squares. Glue the squares in order on a paper strip.

# Riding a Bike

Name_____

Cut out the squares. Glue the squares in order on a paper strip.

# Foot Stomping Fun

## Objectives

*Practice recognizing, ordering, and writing numbers.*

## Materials Needed

❑ worksheets
❑ construction paper
❑ scissors
❑ felt tip marker
❑ large envelopes
❑ pencil

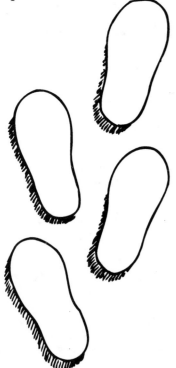

## Setting Up the Station

- Using the Footprint Patterns worksheet on page 71 as a guide, cut 30 footprint shapes out of construction paper.
- Number ten of the footprints from 1 to 10 with a felt tip marker. Put them in a large envelope labeled "1 to 10."
- Use the felt tip marker to number the remaining 20 of the footprints from 1 to 20. Put them in a large envelope labeled "1 to 20."
- Copy the worksheets on pages 72–73.
- Set out the envelopes of numbered footprints, a pencil, and copies of the worksheets.

## The Project

*Explain the following steps to your children.*

1. Select the "1 to 10" or "1 to 20" envelope of footprints.
2. Arrange the footprints in order on the floor. Carefully walk on the footprints while you say the numbers.
3. Mix up the footprints and put them back in the envelope.
4. Select a copy of the Take a Hike: 1 to 10 worksheet and write your name on it.
5. Follow the trail of footprints and write in the missing numbers.
6. If you have time, choose a copy of the Take a Hike: 11 to 20 worksheet and complete it in the same way.
7. Place your worksheets in the designated area.

## Follow-Up

- Call up each child and date his or her worksheets.
- Ask questions to determine the child's level of understanding.
- Further explain or demonstrate concepts as needed.
- Record the child's progress.
- Keep the worksheets in the child's folder or send them home.

# Footprint Patterns

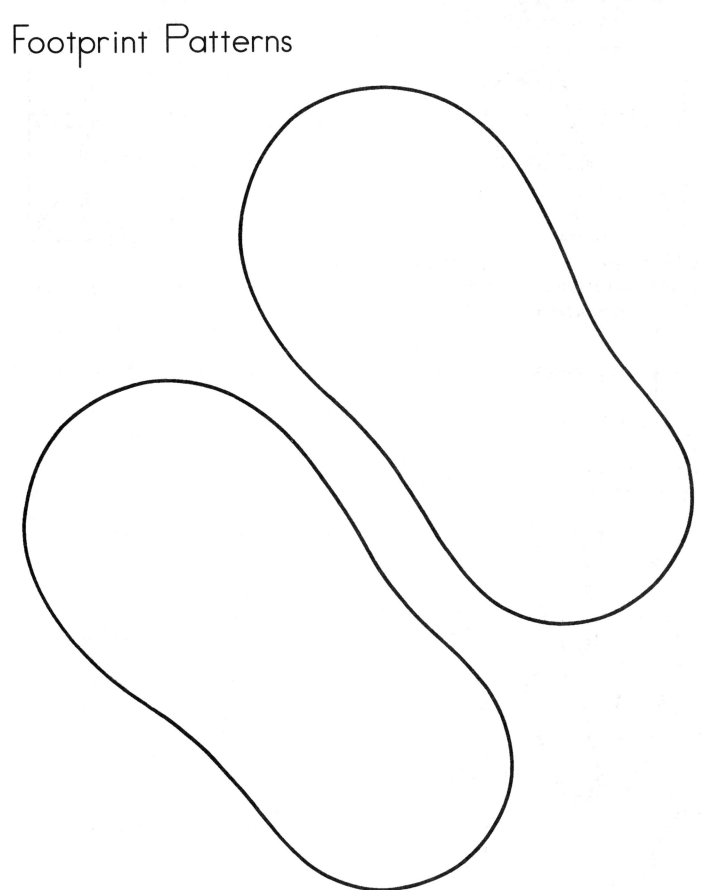

# Take a Hike: 1 to 10

Name_____

Follow the footprints and fill in the missing numbers.

Child understands: ❑ number order, 1 to 10.

# Take a Hike: 11 to 20

Name_____

Follow the footprints and fill in the missing numbers.

Child understands: ❑ number order, 11 to 20.

# Crown of Numbers

## Objectives

*Practice recognizing and ordering numbers.*

## Materials Needed

❑ construction paper
❑ scissors
❑ glue
❑ crayons
❑ pencil
❑ tape
❑ worksheets

## Setting Up the Station

• Make paper crowns by cutting various colors of large sheets of construction paper in half as shown in the illustration.
• Copy the Jewel Patterns worksheets on pages 75–77.
• Set out the crowns, scissors, glue, crayons, a pencil, tape, and copies of the worksheets.

## The Project

*Explain the following steps to your children.*

**1.** Choose one of the paper crowns.

**2.** Decide if you want your crown to have 10, 20, or 30 jewels on it. Select the copies of the corresponding Jewel Patterns worksheets.

**3.** Cut out the jewel shapes.

**4.** Arrange the jewel shapes in number order on your crown and glue them in place.

**5.** Decorate your crown with crayons and write your name on the back.

**6.** Hold the crown around your head and have an adult tape it in place.

**7.** Place your crown in the designated area.

## Follow-Up

• Call up each child and date his or her crown.
• Ask questions to determine the child's level of understanding.
• Further explain or demonstrate concepts as needed.
• Send the crown home.

# Jewel Patterns: 1 to 10

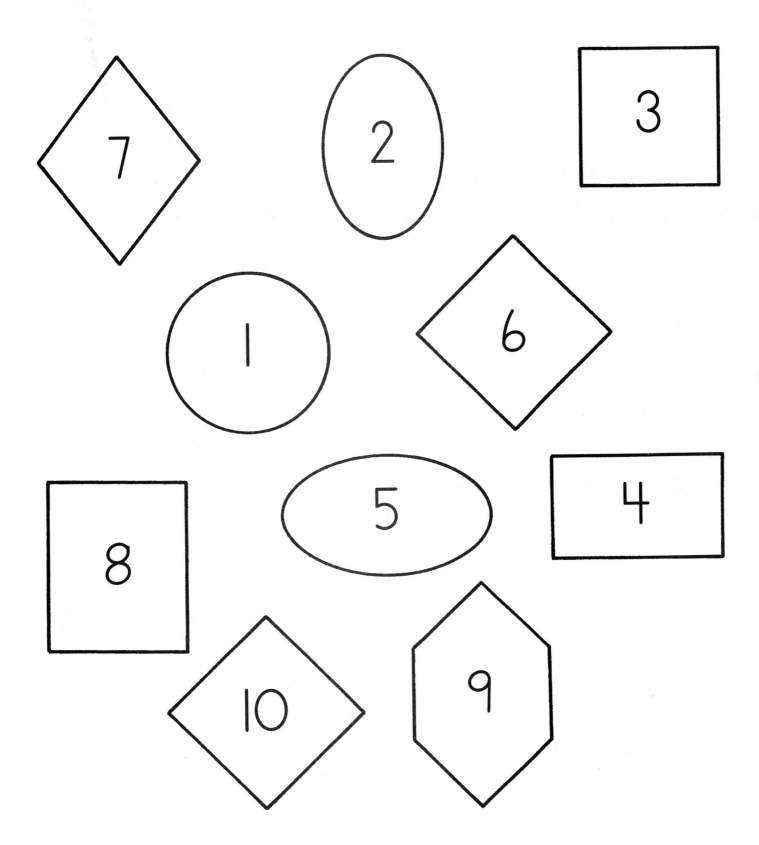

# Jewel Patterns: 11 to 20

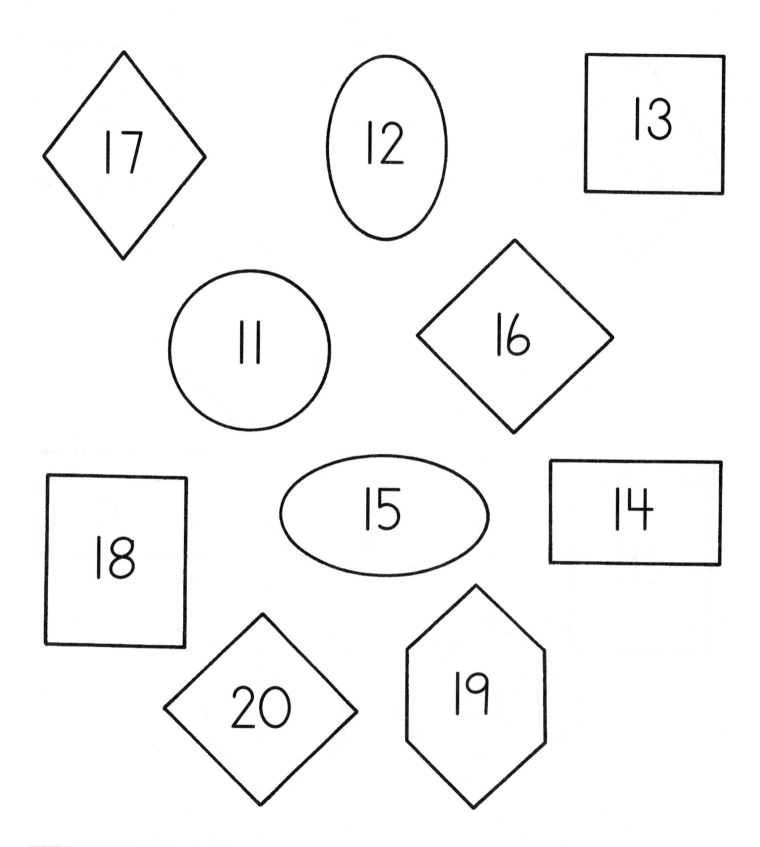

# Jewel Patterns: 21 to 30

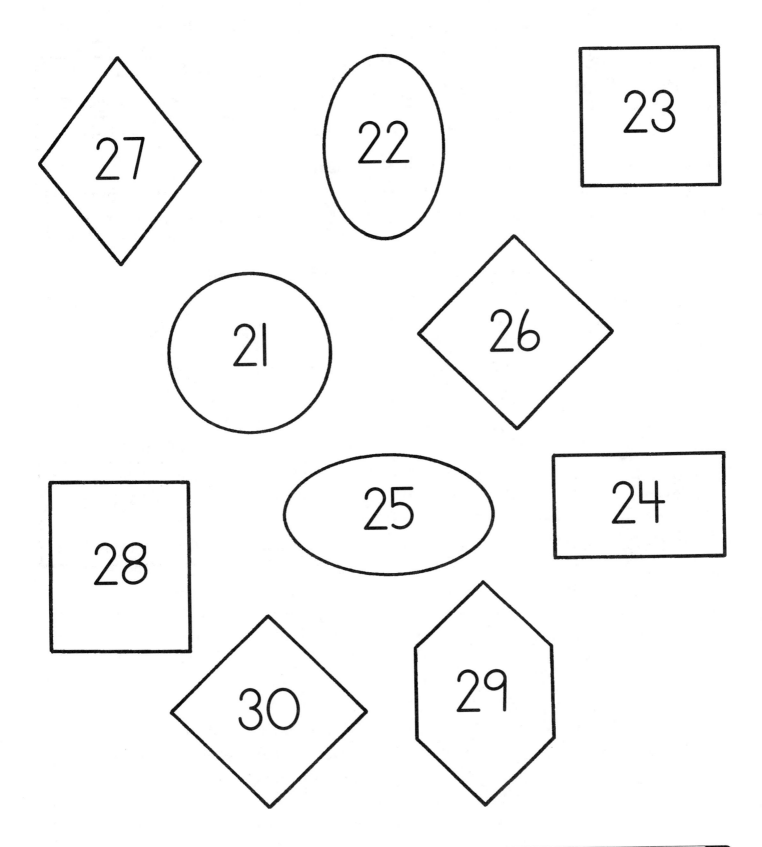

# All Aboard

## Objectives

*Practice recognizing and ordering numbers.*

## Materials Needed

❏ toy train engine and cars
❏ permanent marker
❏ stickers (optional)
❏ box
❏ pencil
❏ scissors
❏ glue
❏ worksheets

## Setting Up the Station

- Collect a toy train set with an engine and ten cars.
- Number the cars from 1 to 10 by using a permanent marker to write on the cars or, if you prefer, to write on stickers and then place them on the cars.
- Put the train engine and numbered train cars in a box.
- Copy the worksheets on pages 79–81.
- Set out the train set, a pencil, scissors, glue, and copies of the worksheets.

## The Project

*Explain the following steps to your children.*

1. Take the toy train engine and cars out of the box.

2. Arrange the train cars in numerical order behind the train engine.

3. Mix up the train cars and engine and put them back in the box.

4. Take a copy of one of the Train Cars worksheets and write your name on it.

5. Help the train pick up its cars in numerical order by drawing a line from car number 1 to car number 2 to car number 3 and so on.

6. If you have time, select a copy of the Toy Trains worksheet and write your name on it. Complete the teddy bear train by counting the teddy bears in each car and gluing them in order behind the teddy bear engine. Do the same for the block engine and block cars.

7. Place your worksheets in the designated area.

## Follow-Up

- Call up each child and date his or her worksheets.
- Ask questions to determine the child's level of understanding.
- Further explain or demonstrate concepts as needed.
- Record the child's progress.
- Keep the worksheets in the child's folder or send them home.

# Train Cars: 1 to 10

Name_____

Connect the train cars in order.

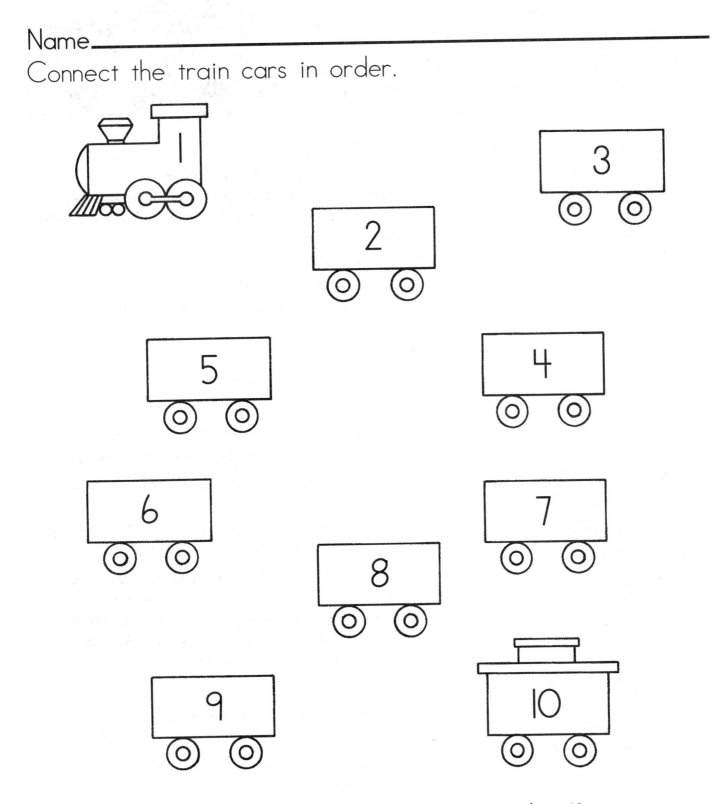

Child understands: ❑ ordering 1 to 10;   ❑ recognizing numerals 1 to 10.

# Train Cars: 11 to 20

Name_____

Connect the train cars in order.

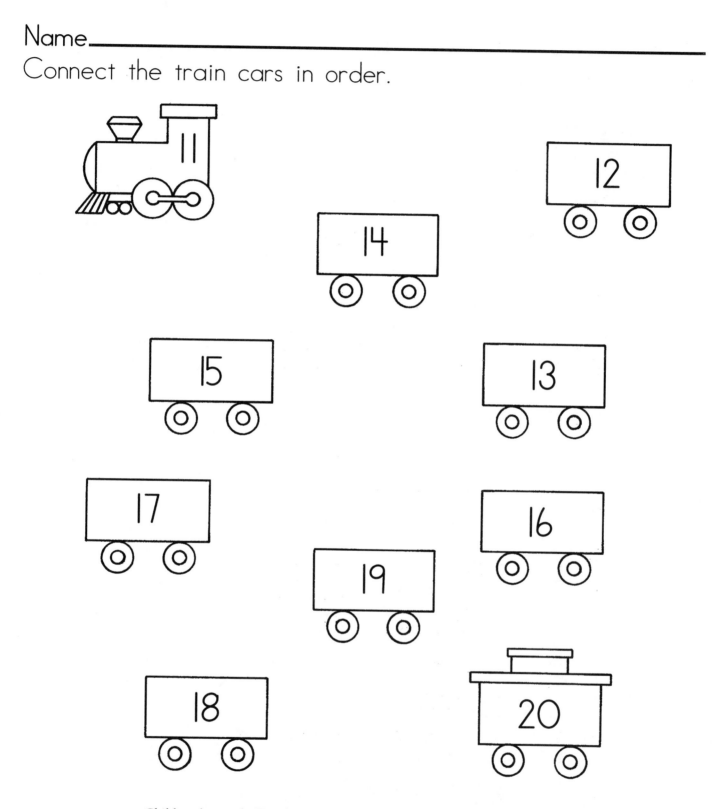

Child understands: ☐ ordering 11 to 20; ☐ recognizing numerals 11 to 20.

# Toy Trains

Name_____

Cut out the teddy bear cars and the block cars. Glue them in order behind the matching engines.

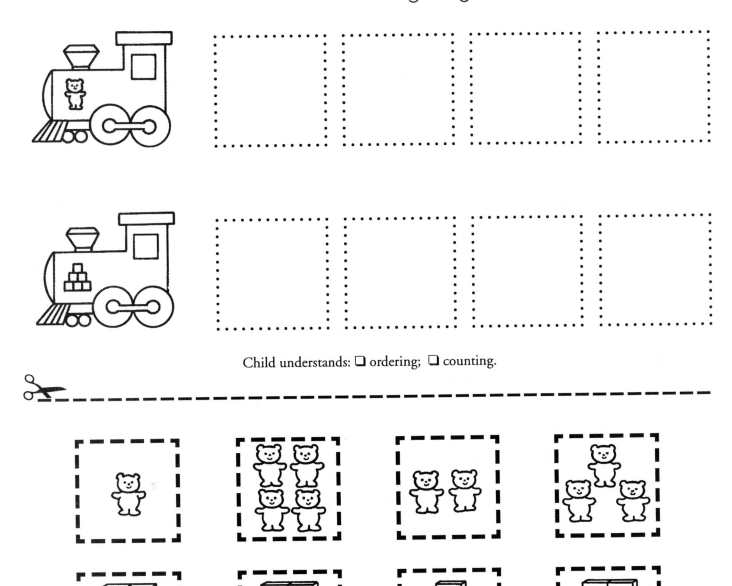

Child understands: ❑ ordering; ❑ counting.

# Graphing

# Graphing Pretzels

## Objectives

*Practice sorting, counting, and graphing.*

## Materials Needed

❑ pretzels
❑ resealable bags
❑ napkins
❑ pencil
❑ crayons
❑ worksheet

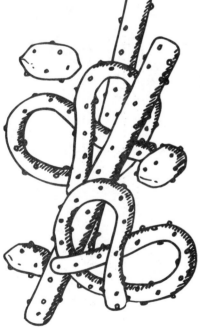

## Setting Up the Station

• Purchase stick pretzels, twist pretzels, and pretzel nuggets.

• Prepare a pretzel bag for each child by placing between one and ten of each kind of pretzel in a resealable bag.

• Copy the worksheet on page 85.

• Set out the pretzel bags, napkins, a pencil, crayons, and copies of the worksheet.

## The Project

*Explain the following steps to your children.*

1. Select one of the bags of pretzels.

2. Open the bag and spread out the pretzels on a napkin.

3. Sort the pretzels by kind: stick pretzels, twist pretzels, and pretzel nuggets.

4. Take a copy of the worksheet and write your name on it.

5. Count how many stick pretzels you have. Find the stick pretzel on the worksheet and color in that many squares.

6. Repeat for the twist pretzels and the pretzel nuggets.

7. Eat your pretzels.

8. Place your worksheet in the designated area.

## Follow-Up

• Call up each child and date his or her worksheet.

• Ask questions to determine the child's level of understanding.

• Further explain or demonstrate concepts as needed.

• Record the child's progress.

• Keep the worksheet in the child's folder or send it home.

# Pretzel Graph

Name_____

Count how many pretzels you have of each kind. Color in that many boxes.

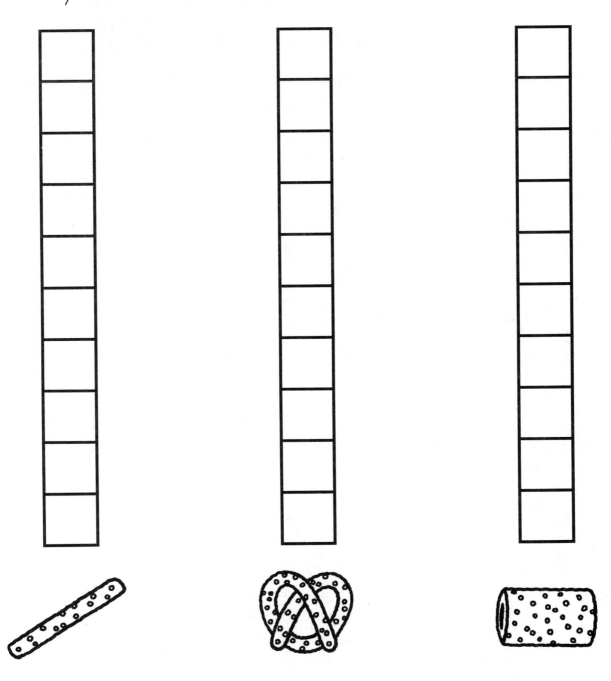

Child understands: ❑ counting;  ❑ one-to-one correspondence; ❑ graphing.

# Candy Graphing

## Objectives

*Practice sorting, counting, and graphing.*

## Materials Needed

❑ colorful candy pieces
❑ resealable bags
❑ napkins
❑ pencil
❑ crayons
❑ worksheets

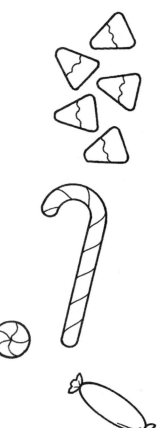

## Setting Up the Station

• Fill resealable bags with some colorful candy pieces. Make sure that there are no more than eight of any one color of candy.

• Copy the worksheets on pages 87–89.

• Set out the bags of candy, napkins, a pencil, crayons, and copies of the worksheets.

## The Project

*Explain the following steps to your children.*

1. Select a bag of candy.

2. Empty the candy onto a napkin.

3. Sort the candy by color.

4. Take a copy of the Colorful Candy Graph worksheet and write your name on it.

5. Count one of the colors of candy. Use a matching-colored crayon to fill in that same number of candies on one of the worksheet rows.

6. Repeat with the remaining colors of candy.

7. Eat your candy.

8. If you have time, select a copy of the Count and Graph worksheet and write your name on it. For each row, count the candies in the first box of each row and color that number of squares.

9. If you still have time, select a copy of the Which Is More? worksheet and write your name on it. For each box, circle the candy with the most shown on the graph.

10. Place your worksheets in the designated area.

## Follow-Up

• Call up each child and date his or her worksheets.

• Ask questions to determine the child's level of understanding.

• Further explain or demonstrate concepts as needed.

• Record the child's progress.

• Keep the worksheets in the child's folder or send them home.

# Colorful Candy Graph

Name_____

Color the number of candies to match yours.

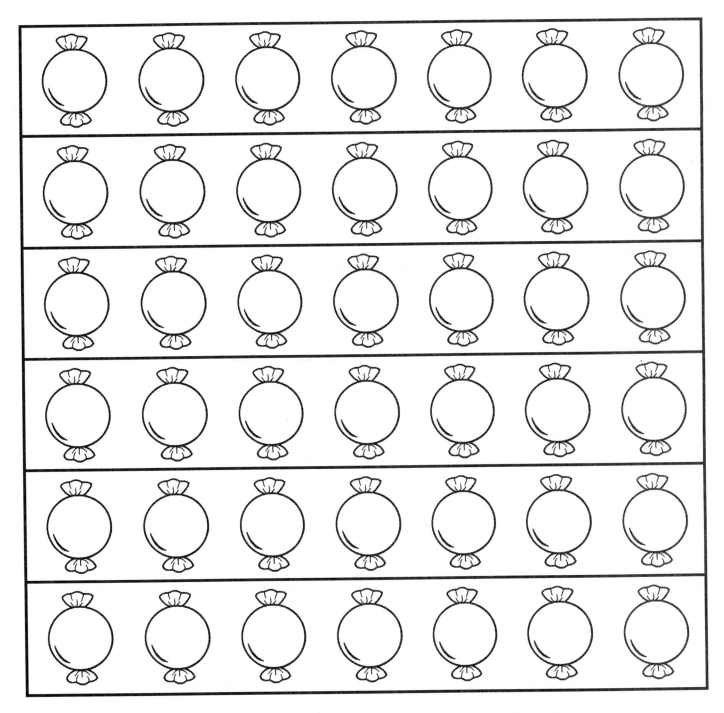

Child understands: ❑ counting; ❑ one-to-one correspondence; ❑ graphing.

# Count and Graph

Name_____

For each row, count the candies and color in that number of squares.

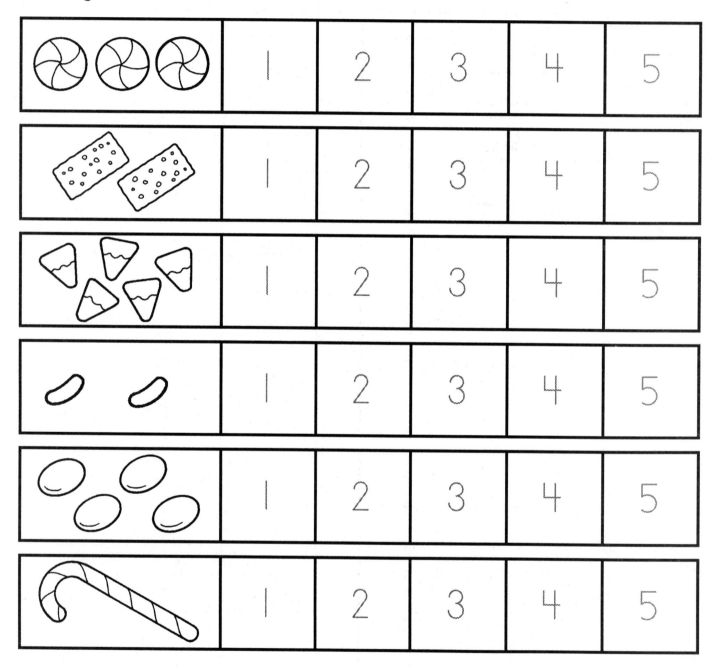

Child understands: ❑ counting; ❑ one-to-one correspondence; ❑ graphing.

# Which Is More?

Name _____

For each box, circle the candy with the most shown.

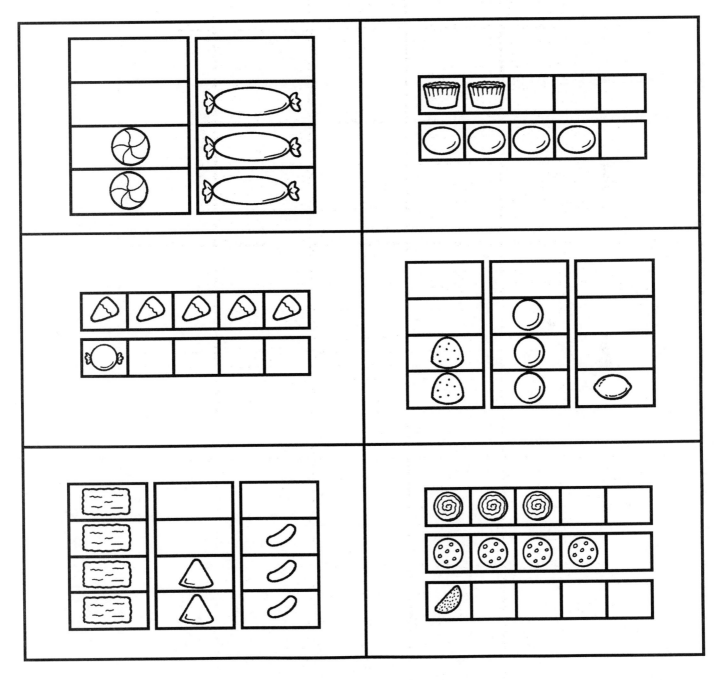

Child understands: ❏ more and less; ❏ graphing.

# Buggy Graphs

## Objectives

*Practice sorting, reading and writing numerals, counting, and graphing.*

## Materials Needed

❑ worksheets
❑ scissors
❑ box
❑ pencil
❑ glue

## Setting Up the Station

- Copy the Bug Patterns worksheet on page 91. (You will need about one copy for each child.)
- Cut out the individual bug squares.
- Place the bug squares in a box.
- Copy the worksheets on pages 92–93.
- Set out the box of bug squares, a pencil, glue, and copies of the worksheets.

## The Project

*Explain the following steps to your children.*

1. Carefully sort a handful of the bug squares.
2. Take a copy of the Graphing Bugs worksheet and write your name on it.
3. Look at the number and the kind bug at the bottom of the first column.
4. Count out that many of that kind of bug (four ladybugs). Starting from the bottom, glue each ladybug in one of the boxes in the column.
5. Repeat with the remaining columns.
6. If you have time, take a copy of the Make a Graph worksheet and write your name on it. Complete the worksheet by gluing several of the same kind of bug in the boxes in the first column. Write the number of bugs on the line below. Repeat for the remaining columns.
7. Place your worksheets in the designated area.

## Follow-Up

- Call up each child and date his or her worksheets.
- Ask questions to determine the child's level of understanding.
- Further explain or demonstrate concepts as needed.
- Record the child's progress.
- Keep the worksheets in the child's folder or send them home.

# Bug Patterns

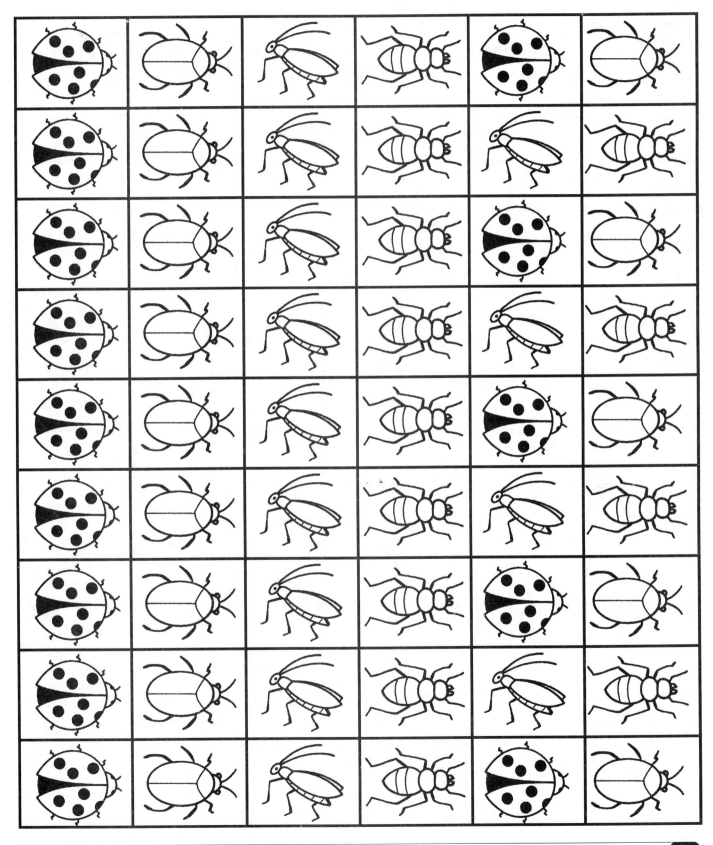

# Graphing Bugs

Name_____

For each column, say the number. Glue that many of that kind of bug in the boxes.

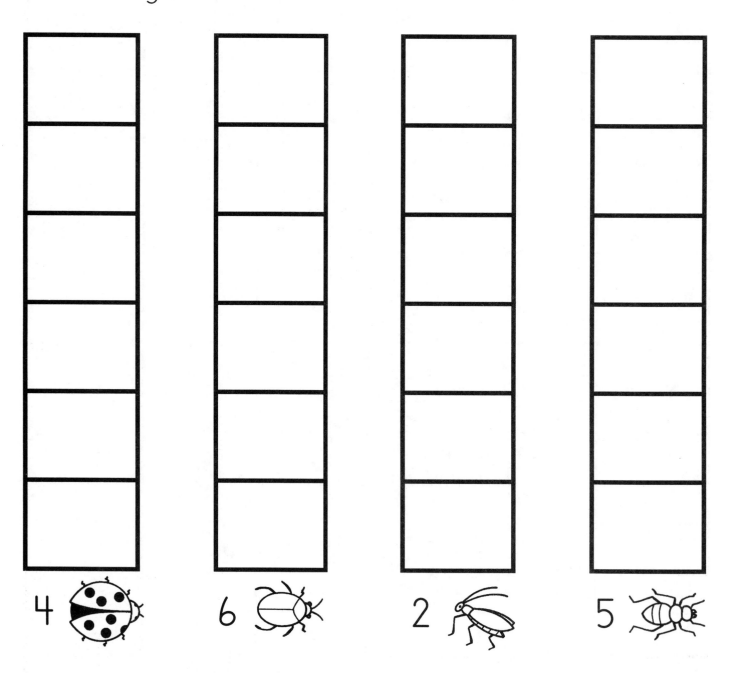

Child understands: ❏ reading numerals; ❏ counting; ❏ graphing.

# Make a Graph

Name _____

Glue several of a different kind of bug in each column.
Count the bugs. Write the number on the line below.

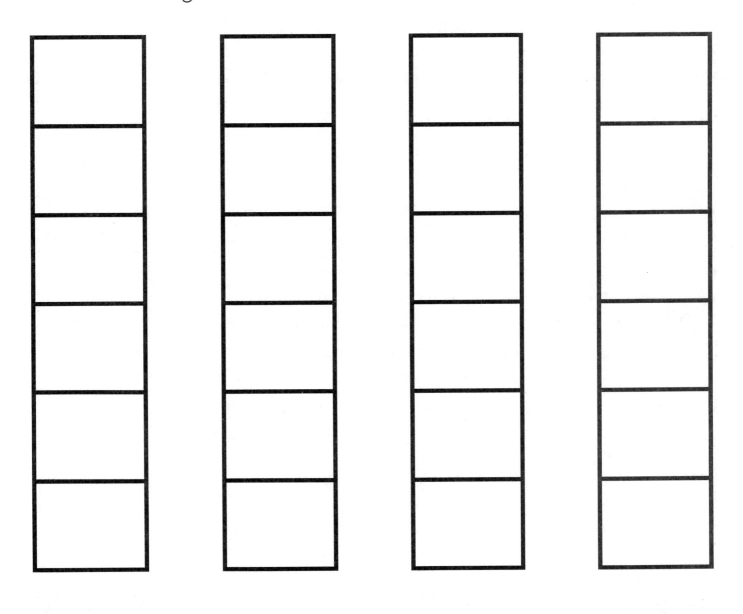

Child understands: ❑ graphing; ❑ counting; ❑ writing numerals.

# Adding and Subtracting

# Fruity Math

## Objectives

*Practice counting and writing numerals and develop adding skills.*

## Materials Needed

❑ orange
❑ banana
❑ modeling clay
❑ containers with lids
❑ pencil
❑ worksheet

## Setting Up the Station

• Arrange a fresh orange and banana on the table at the station.
• Purchase yellow- and orange-colored modeling clay. Put them in separate containers with lids.
• Copy the worksheet on page 97.
• Set out the containers of modeling clay, a pencil, and copies of the worksheet.

## The Project

*Explain the following steps to your children.*

1. Look at the orange and the banana. Notice their shapes.
2. Make tiny orange and banana shapes out of the orange and yellow modeling clay.
3. Put all of your clay oranges together and count them.
4. Put all of your clay bananas together and count them.
5. Use the clay oranges and bananas to make addition problems. For example, set out two oranges and three bananas. How many fruit pieces do you have in all? Try making several different problems to solve.
6. Take a copy of the worksheet and write your name on it.
7. Look at the first problem. Count all the fruit and write that number on the line. Repeat for the remaining problems. Use your clay fruit as counters, if you wish.
8. Squish the clay and put each color into its own container.
9. Place your worksheet in the designated area.

## Follow-Up

• Call up each child and date his or her worksheet.
• Ask questions to determine the child's level of understanding.
• Further explain or demonstrate concepts as needed.
• Record the child's progress.
• Keep the worksheet in the child's folder or send it home.

# Oranges and Bananas

Name_____

For each row, add the fruits together. Write the number on the line.

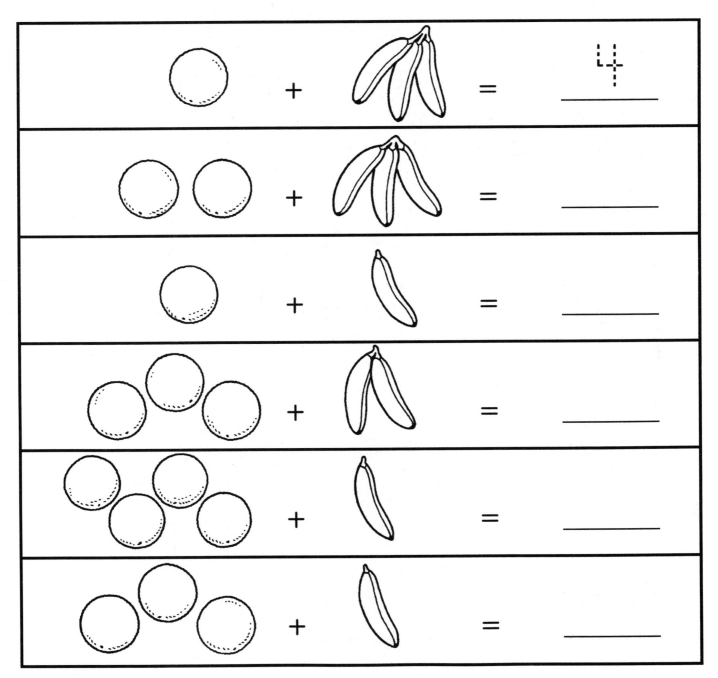

Child understands: ❏ counting; ❏ adding up to 5; ❏ writing numerals.

# Adding Clouds

## Objectives

*Practice counting and writing numerals and develop adding skills.*

## Materials Needed

- ❏ cotton balls
- ❏ basket
- ❏ construction paper
- ❏ scissors
- ❏ felt tip marker
- ❏ pencil
- ❏ worksheets

## Setting Up the Station

- Place two or three handfuls of cotton balls in a basket.
- Cut cloud shapes out of construction paper and write a simple addition problem on each one.
- Copy the worksheets on pages 99–101.
- Set out the basket of cotton balls, paper clouds, a pencil, and copies of the worksheets.

## The Project

*Explain the following steps to your children.*

1. Choose one of the paper clouds.

2. Use the cotton balls to help you solve the problem on the cloud.

3. Repeat with the other clouds.

4. Put the cotton balls back in the basket and set the paper clouds aside.

5. Select a copy of one of the Add the Clouds worksheets and write your name on it.

6. On the worksheet, count the total number of clouds in each problem and write the number on the line.

7. If there is time, write your name on a copy of the Find the Cloud worksheet. Complete it by solving each problem and drawing a line to the cloud with the correct answer.

8. Place your worksheets in the designated area.

## Follow-Up

- Call up each child and date his or her worksheets.
- Ask questions to determine the child's level of understanding.
- Further explain or demonstrate concepts as needed.
- Record the child's progress.
- Keep the worksheets in the child's folder or send them home.

# Add the Clouds: Up to 5

Name_____

For each row, count the clouds. Write the number on the line.

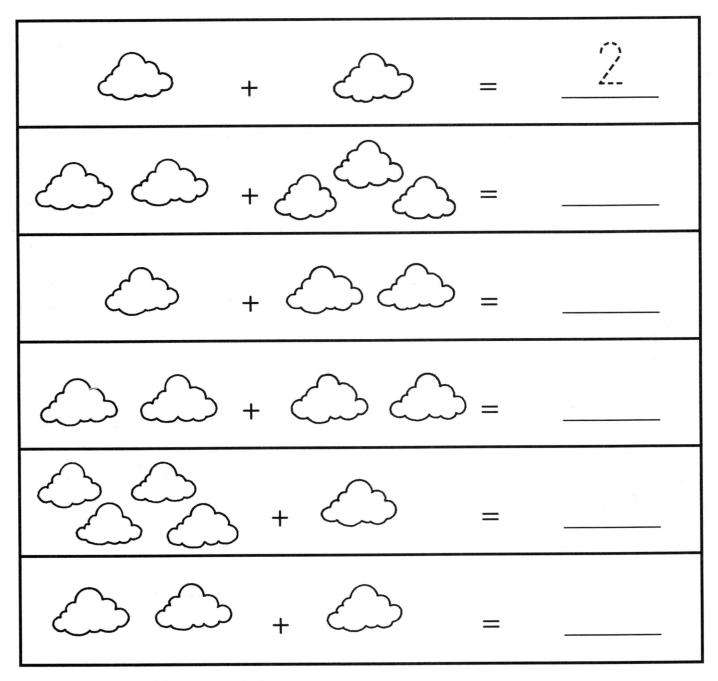

Child understands: ❏ counting; ❏ adding up to 5; ❏ writing numerals.

# Add the Clouds: Up to 10

Name_____

For each row, count the clouds. Write the number on the
line.

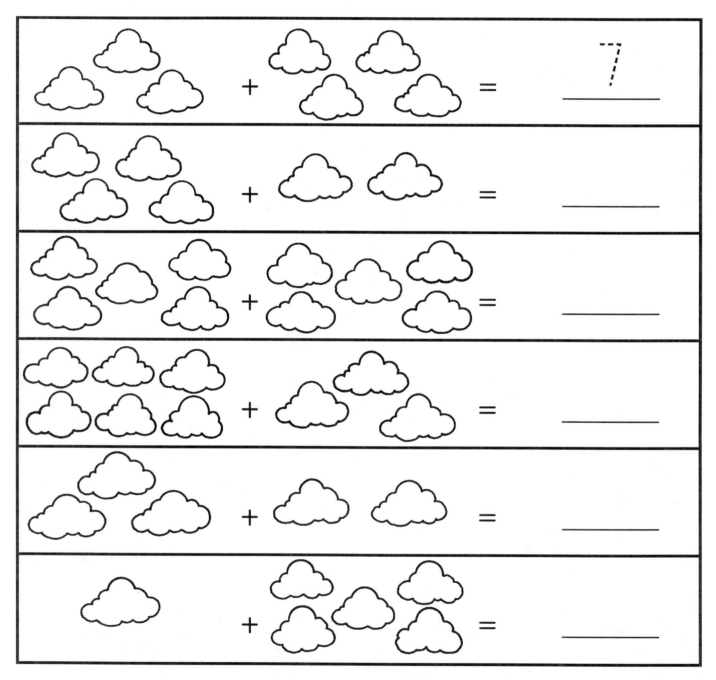

Child understands: ❑ counting;  ❑ adding up to 10; ❑ writing numerals.

# Find the Cloud

Name_____

Solve each problem. Draw a line to the cloud with the correct answer.

2 + 1 =

1 + 3 =

1 + 1 =

3 + 2 =

0 + 1 =

(Cloud with 4)

(Cloud with 5)

(Cloud with 3)

(Cloud with 1)

(Cloud with 2)

Child understands: ❏ adding up to 5; ❏ reading numerals; ❏ writing numerals.

# Bean Count

## Objectives

*Practice counting and develop adding skills.*

## Materials Needed

❑ dry lima beans

❑ bowl

❑ index cards

❑ felt tip marker

❑ pencil

❑ worksheets

## Setting Up the Station

• Open a bag of dry lima beans and put them in a bowl.

• Purchase or find about five large index cards.

• Turn each index card into a blank problem card by drawing on a problem that looks like this: ☐ + ☐ = ☐.

• Copy the worksheets on pages 103–105.

• Set out the bowl of lima beans, blank problem cards, a pencil, and copies of the worksheets.

## The Project

*Explain the following steps to your children.*

1. Use the lima beans to make your own addition problems.

2. Select one of the blank problem cards; count out some lima beans for the first box and some lima beans for the second box. Count the beans in both boxes and put that many lima beans in the answer box. Make several different problems.

3. Select one of the Add Them worksheets and write your name on it.

4. For each problem, look at the first box and draw in the correct number of beans. Do the same for the second box.

5. Add the beans in the two boxes together and draw that number of beans in the third box.

6. If time allows, select additional worksheets to complete.

7. Place your worksheets in the designated area.

## Follow-Up

• Call up each child and date his or her worksheets.

• Ask questions to determine the child's level of understanding.

• Further explain or demonstrate concepts as needed.

• Record the child's progress.

• Keep the worksheets in the child's folder or send them home.

# Add Them: Up to 5

Name_____

Draw the correct number of beans in each box.

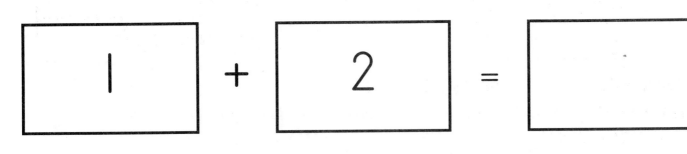

| 1 | + | 2 | = | |

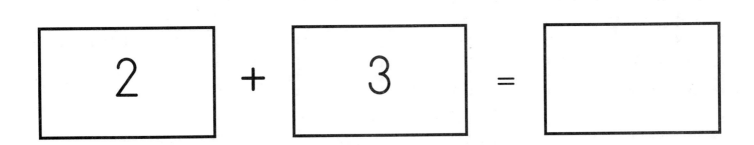

| 2 | + | 3 | = | |

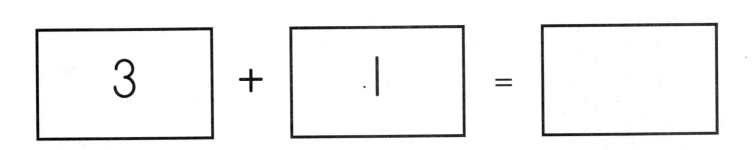

| 3 | + | 1 | = | |

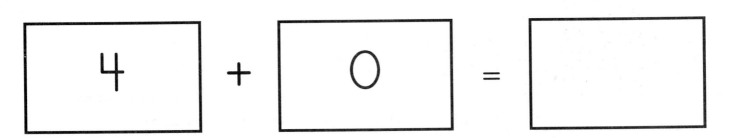

| 4 | + | 0 | = | |

Child understands: ❏ counting; ❏ one-to-one correspondence; ❏ adding up to 5.

# Add Them: Up to 7

Name_____

Draw the correct number of beans in each box.

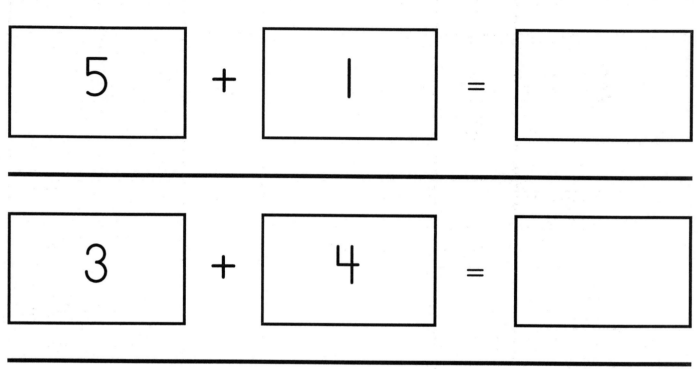

| 5 | + | 1 | = | |

| 3 | + | 4 | = | |

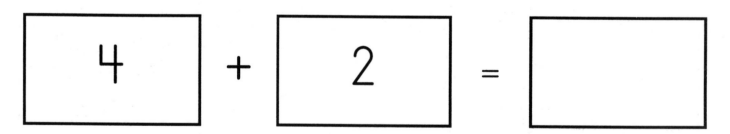

| 4 | + | 2 | = | |

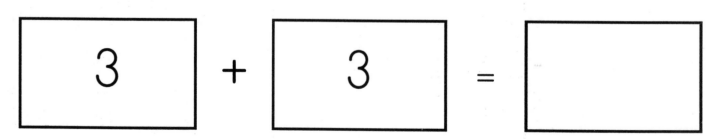

| 3 | + | 3 | = | |

Child understands: ❑ counting;  ❑ one-to-one correspondence;  ❑ adding up to 7.

# Add Them: Up to 10

Name_____

Draw the correct number of beans in each box.

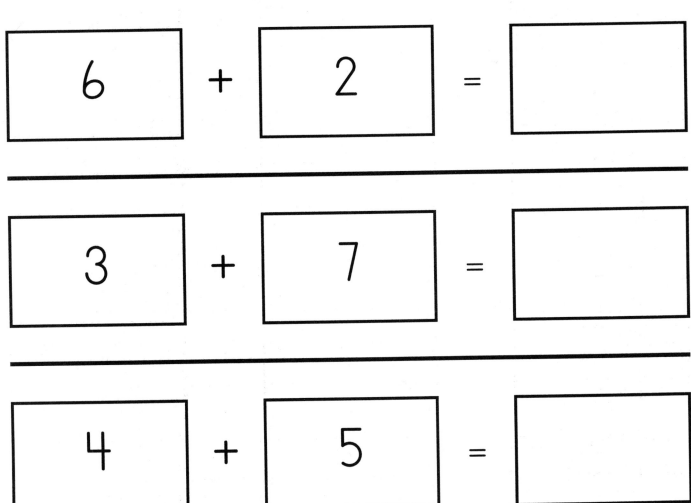

| 6 | + | 2 | = | |
| 3 | + | 7 | = | |
| 4 | + | 5 | = | |
| 9 | + | 1 | = | |

Child understands: ❑ counting; ❑ one-to-one correspondence; ❑ adding up to 10.

# Take Away Beads

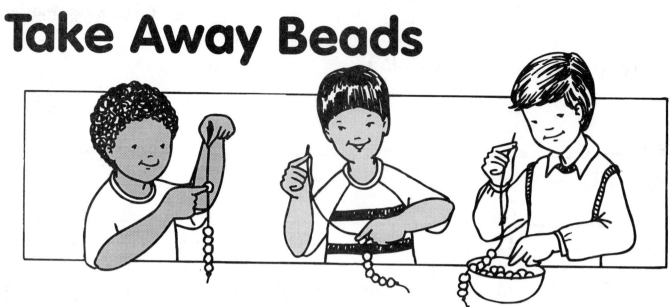

## Objectives

*Practice counting and writing numerals and develop subtracting skills.*

## Materials Needed

❏ large wooden beads
❏ shoelace
❏ basket
❏ pencil
❏ worksheets

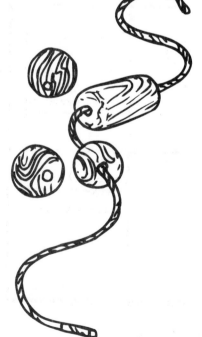

## Setting Up the Station

• Collect a variety of large wooden beads.
• Find an extra shoelace. Tie a large knot in one end.
• Put the wooden beads and shoelace in a basket.
• Copy the worksheets on pages 107–109.
• Set out the basket of beads and shoelace, a pencil, and copies of the worksheets.

## The Project

*Explain the following steps to your children.*

1. String several beads on the shoelace. Count the beads. Take one bead off the shoelace. How many are left? Take two beads off. Now how many are left?

2. Repeat several times, stringing beads on and taking them off.

3. Select one of the Leftover Beads worksheets and write your name on it.

4. For each problem, cross off the number of beads that are being taken away.

5. Count how many beads are left and write that number in the box.

6. If time allows, select additional worksheets to complete.

7. Place your worksheets in the designated area.

## Follow-Up

• Call up each child and date his or her worksheets.
• Ask questions to determine the child's level of understanding.
• Further explain or demonstrate concepts as needed.
• Record the child's progress.
• Keep the worksheets in the child's folder or send them home.

# Leftover Beads: 5 or Less

Name_____

Cross off the number of beads taken away. Write the number of beads left over.

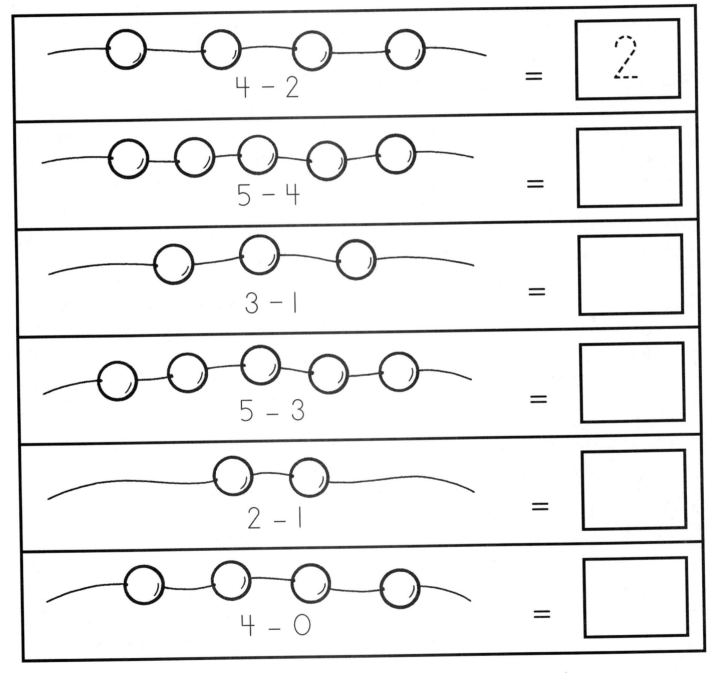

Child understands: ❑ counting; ❑ subtracting from 5; ❑ writing numerals.

*Adding and Subtracting* • Calculation Station

# Leftover Beads: 7 or Less

Name_____

Cross off the number of beads taken away. Write the number of beads left over.

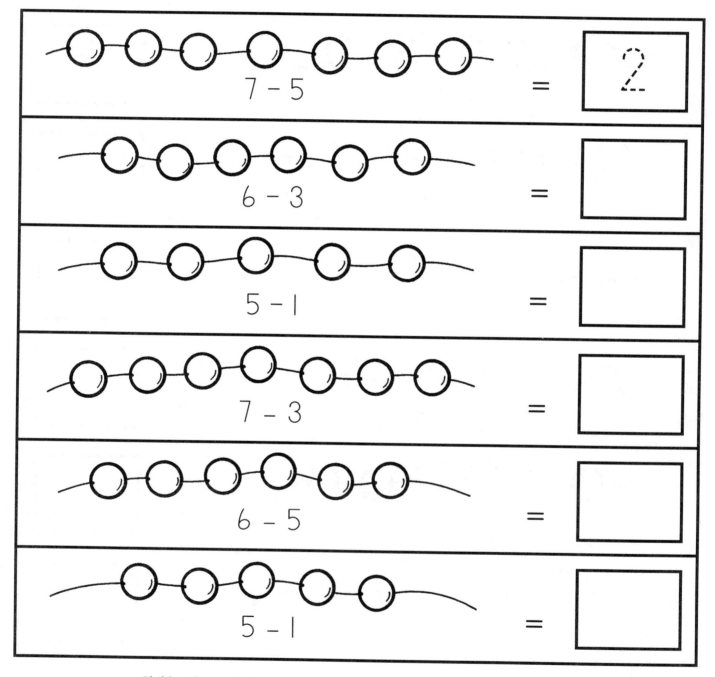

Child understands: ☐ counting; ☐ subtracting from 7; ☐ writing numerals.

# Leftover Beads: 10 or Less

Name_____

Cross off the number of beads taken away. Write the number of beads left over.

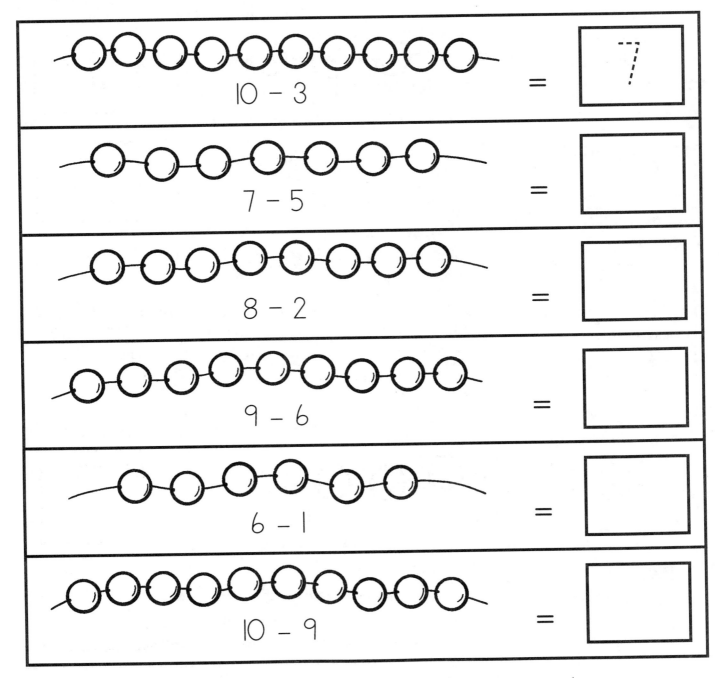

# Sailing Away

## Objectives

*Practice counting, identifying and writing numerals, and one-to-one correspondence, and develop subtraction skills.*

## Materials Needed

- ❏ worksheets
- ❏ scissors
- ❏ felt tip marker
- ❏ craft sticks
- ❏ glue
- ❏ modeling dough
- ❏ paper cups
- ❏ crackers
- ❏ bowl
- ❏ pencil

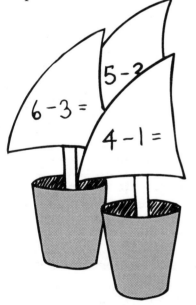

## Setting Up the Station

- Make several copies of the Subtraction Sail Patterns worksheet on page 111.
- Cut out the sails and write a different subtraction problem on each one. Select problems that your children are currently working on.
- Glue each sail to a craft stick.
- For each sail, put a small amount of modeling dough in the bottom of a separate paper cup. Stick the craft stick into the dough so the sail shows above the cup, to make a "subtraction boat."
- Purchase goldfish-shaped crackers or oyster crackers. Put the crackers in a bowl.
- Copy the worksheets on pages 112 and 113.
- Set out the subtraction boats, the bowl of crackers, a pencil, and copies of the worksheets.

## The Project

*Explain the following steps to your children.*

1. Select one of the subtraction boats.

2. Look at the problem on the sail. Identify the first number and count out that many crackers. Look at the second number and take away (and eat!) that many crackers. Count how many crackers you have left. Then eat them, too.

3. Repeat for other subtraction boats.

4. Select one of the Subtraction Boats worksheets and write your name on it.

5. Complete the worksheet by solving the problem written on each sail and writing the answer on the boat.

6. If time allows, do the other Subtraction Boat worksheet.

7. Place your worksheets in the designated area.

## Follow-Up

- Call up each child and date his or her worksheets.
- Ask questions to determine the child's level of understanding.
- Further explain or demonstrate concepts as needed.
- Record the child's progress.
- Keep the worksheets in the child's folder or send them home.

# Subtraction Sail Patterns

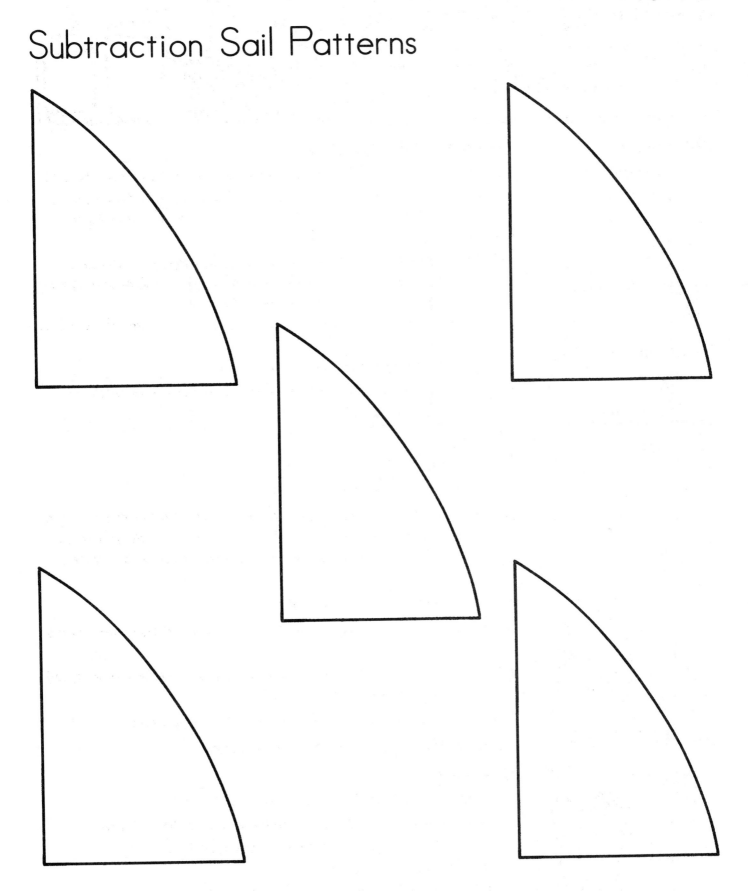

# Subtraction Boats: 5 or Less

Name_____

For each boat, solve the subtraction problem on the sail.
Write the answer in the boat.

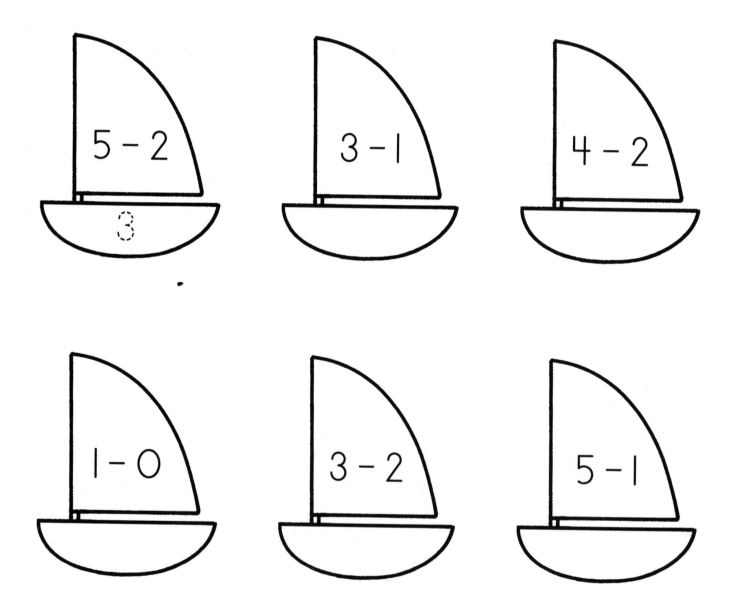

Child understands: ❑ subtracting from 5; ❑ identifying numerals; ❑ writing numerals; ❑ one-to-one correspondence.

# Subtraction Boats: 10 or Less

Name_____

For each boat, solve the subtraction problem on the sail.
Write the answer in the boat.

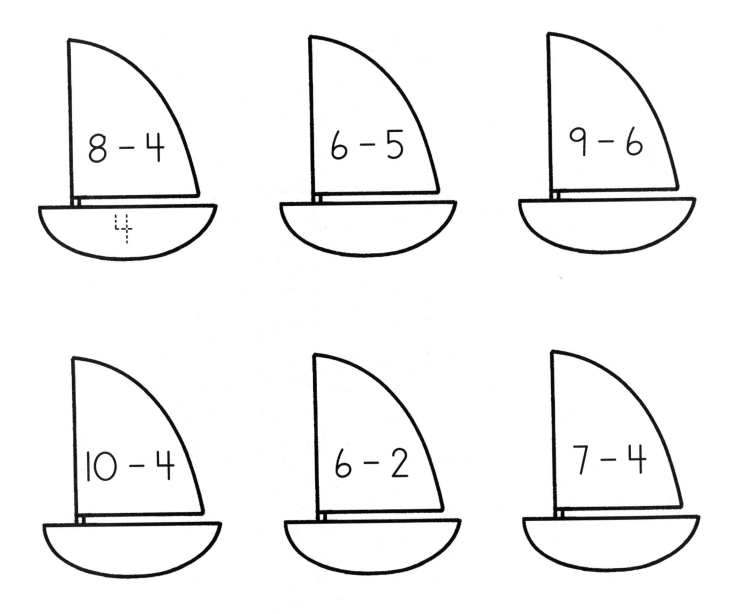

Child understands: ❏ subtracting from 10; ❏ identifying numerals; ❏ writing numerals; ❏ one-to-one correspondence.

# Number Fun

# Dough Numbers

## Objectives

*Practice forming numerals.*

## Materials Needed

☐ modeling dough
☐ resealable plastic bags
☐ clear self-stick paper
☐ scissors
☐ worksheets

## Setting Up the Station

- Purchase or make modeling dough.
- Find a resealable plastic bag for each of your children. Put a handful of the modeling dough in each bag.
- Copy the worksheets on pages 117–119.
- Cover one copy of each worksheet with clear self-stick paper (or have them laminated).
- Prepare worksheet packets for the children to take home by stapling one of each kind of worksheet together.
- Set out the bags of modeling dough, the covered worksheets, and the worksheet packets.

## The Project

*Explain the following steps to your children.*

1. Select one of the prepared Numbers worksheets and a bag of modeling dough.
2. Roll the dough into "snakes."
3. Shape the dough around the outline of the numbers on the worksheet.
4. Roll the dough back together and put it back into the bag.
5. If you have time, select another Numbers worksheet to complete.
6. Take one of the modeling dough bags and a set of Numbers worksheets home.

## Follow-Up

- Call up each child and have him or her demonstrate making a modeling dough numeral.
- Ask questions to determine the child's level of understanding.
- Further explain or demonstrate concepts as needed.

# Numbers: 1 to 5

# Numbers: 6 to 10

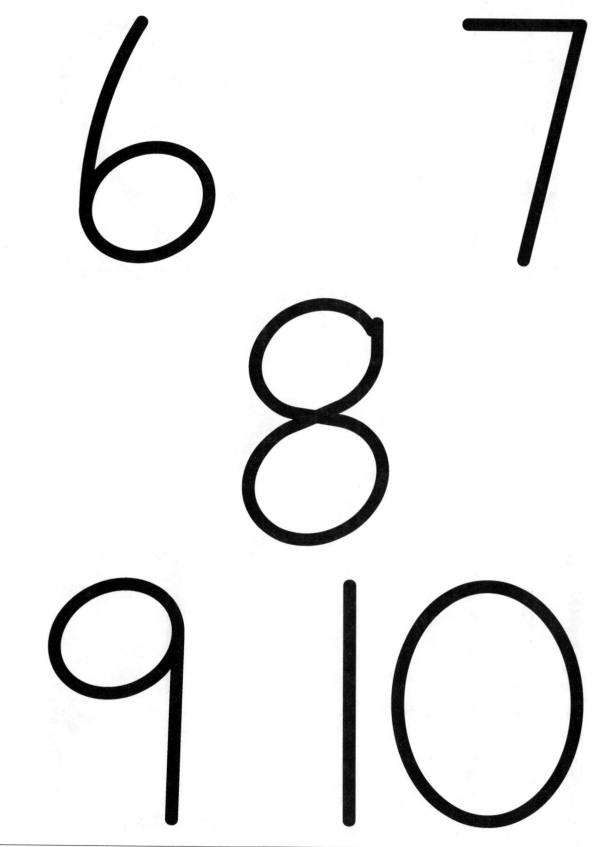

# Numbers: 11 to 15

# Calendar Numbers

## Objectives

*Practice identifying and writing numerals and learn about calendars.*

## Materials Needed

❑ worksheets
❑ felt tip marker
❑ construction paper
❑ glue
❑ pencil
❑ crayons

## Setting Up the Station

- Copy the worksheets on pages 121–123.
- Take a copy of the Calendar: Month and the Calendar: Days worksheets and fill them in for the current month. Glue them to a sheet of 12-by-18-inch construction paper, and hang the completed calendar in the station.
- Set out a pencil, crayons, glue, large sheets of construction paper, and copies of the worksheets.

## The Project

*Explain the following steps to your children.*

1. Select a copy of the Calendar: Month and the Calendar: Days worksheets.
2. Following the example in the station, write the numbers in the boxes on your Calendar: Days worksheet.
3. Following the example in the station, write the month and year on your Calendar: Month worksheet. Write your name on the line at the bottom.
4. Draw a picture in the box on the Calendar: Month worksheet.
5. Select a large sheet of construction paper and fold it in half. Unfold the paper and smooth out the fold.
6. Glue your Calendar: Month worksheet on the top half of the construction paper and your Calendar: Days worksheet on the bottom half.
7. If you have time, complete the Special Dates worksheet by looking at a symbol in the list, finding it on the calendar, and writing the date on the line next to it.
8. Place your calendar in the designated area.

## Follow-Up

- Call up each child and date his or her calendar and worksheet.
- Ask questions to determine the child's level of understanding.
- Further explain or demonstrate concepts as needed.
- Record the child's progress.
- Keep the calendar and worksheet in the child's folder or send them home.

# Calendar: Month

Year

Month

Name

# Calendar: Days

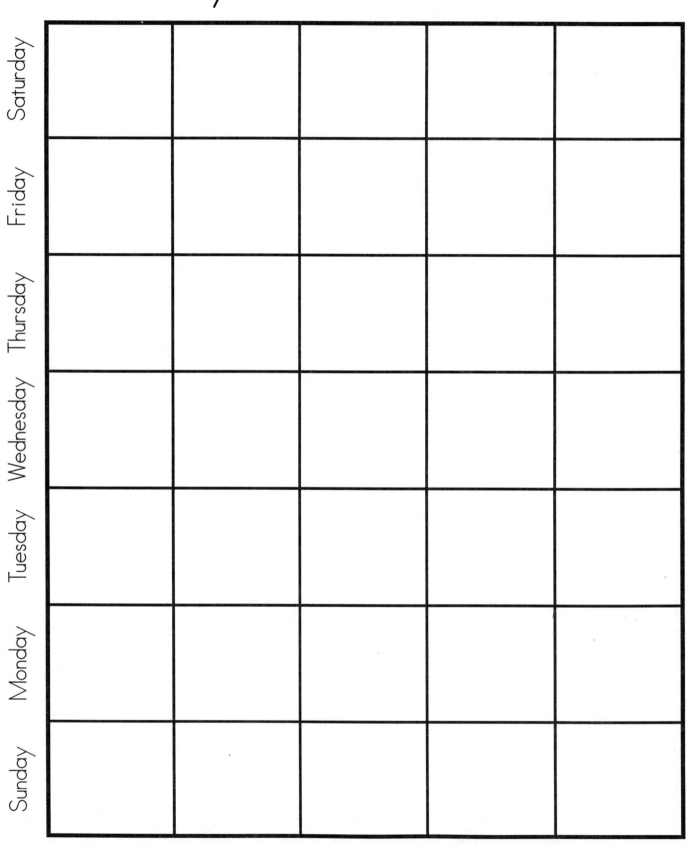

# Special Dates

Name_____

Find each symbol on the calendar. Write the symbol's date on the line beside it.

 _____         _____

🌈 _____        ♡ _____

🦋 _____        ☆ _____

| | | 1 | 2 🧳 | 3 | 4 | 5 |
|---|---|---|---|---|---|---|
| 6 | 7 🌈 | 8 | 9 | 10 ♡ | 11 | 12 |
| 13 | 14 | 15 | 16 | 17 | 18 | 19 ☆ |
| 20 🌷 | 21 | 22 | 23 | 24 | 25 | 26 |
| 27 | 28 | 29 🦋 | 30 | 31 | | |

# Math Puppet

## Objectives

*Identify numerals and math symbols.*

## Materials Needed

❑ paper sacks
❑ pencil
❑ scissors
❑ glue
❑ crayons
❑ worksheet

## Setting Up the Station

• Collect or purchase enough paper lunch sacks for each child to have one.
• Copy the Math Patterns worksheet on page 125.
• Set out the paper sacks, a pencil, scissors, glue, crayons, and copies of the worksheet.

## The Project

*Explain the following steps to your children.*

1. Select one of the paper sacks and write your name on the back of it.
2. Take a copy of the Math Patterns worksheet.
3. Cut out the number and symbol shapes on the worksheet.
4. Arrange the shapes on your paper bag to make a puppet's eyes, nose, mouth, and other features. Glue the shapes in place.
5. Add any other details you would like with crayons.
6. Place your completed puppet in the designated area.

## Follow-Up

• Call up each child and date his or her puppet.
• Ask questions to determine the child's level of understanding.
• Further explain or demonstrate concepts as needed.
• Keep the puppet in the child's folder or send it home.

# Math Patterns

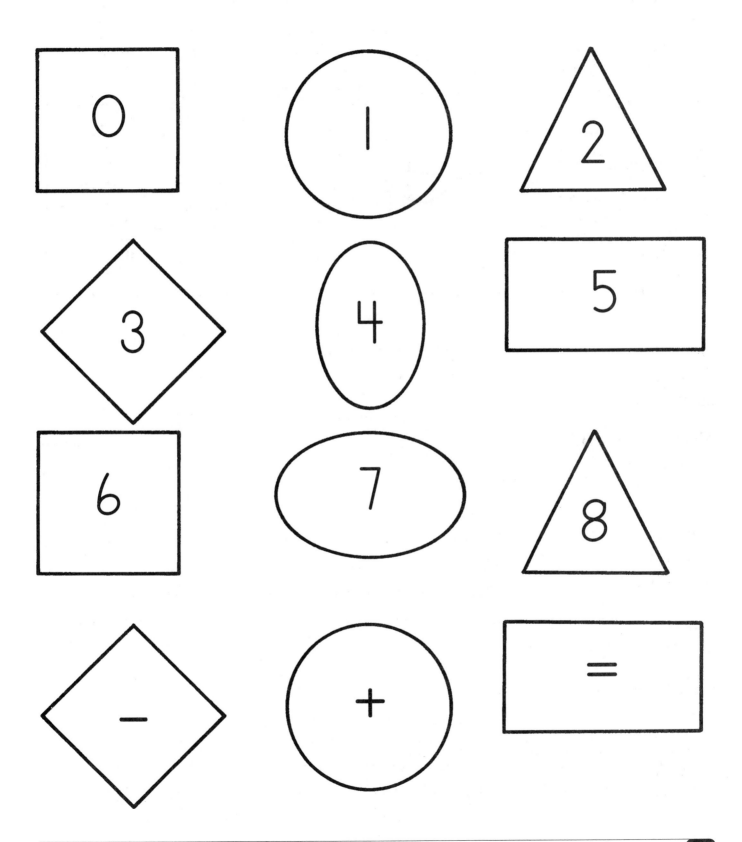

# More
# Math Fun

# How Long?

## Objectives

*Develop an understanding of long and short, and beginning measurement skills.*

## Materials Needed

❑ ribbon
❑ scissors
❑ basket
❑ pencil
❑ worksheet

## Setting Up the Station

• Make a set of measuring ribbons by cutting 1-inch-wide ribbon into seven different lengths. (Make sure that the lengths match the lengths of at least some objects in your room, such as a desk, a chair, or a door.)
• Put the ribbons in a basket.
• Copy the worksheet on page 129.
• Set out the basket of ribbons, a pencil, and copies of the worksheet.

## The Project

*Explain the following steps to your children.*

1. Take the ribbons out of the basket. Arrange them in order from shortest to longest.

2. Select one of the ribbons. Try to find something in the room that is the same length as the ribbon.

3. Put the ribbons back in the basket.

4. Take a copy of the worksheet and write your name on it.

5. Circle the longest line in each of the boxes at the top. Circle the shortest line in each of the boxes at the bottom.

6. Place your completed worksheet in the designated area.

## Follow-Up

• Call up each child and date his or her worksheet.
• Ask questions to determine the child's level of understanding.
• Further explain or demonstrate concepts as needed.
• Record the child's progress.
• Keep the worksheet in the child's folder or send it home.

# Long and Short

Name_____

Circle the longest line.

Circle the shortest line.

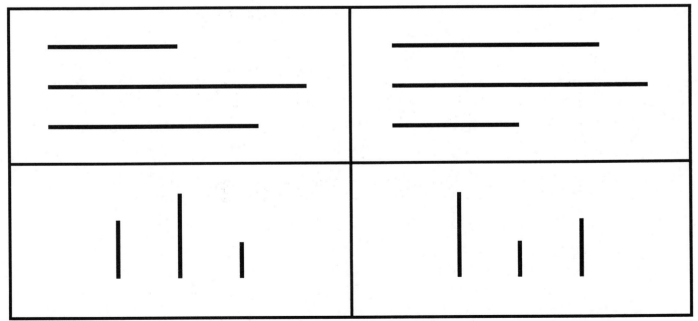

Child understands: ❑ longest; ❑ shortest.

# Tick Tock

## Objectives

*Practice reading a clock with hands.*

## Materials Needed

❑ paper plates
❑ brass paper fasteners
❑ scissors
❑ glue
❑ pencil
❑ worksheets

## Setting Up the Station

• Find or purchase a paper plate and a brass paper fastener for each child.
• Copy the worksheets on pages 131–133.
• Set out the paper plates, brass paper fasteners, scissors, glue, a pencil, and copies of the worksheets.

## The Project

*Explain the following steps to your children.*

1. Select a copy of the Clock Pattern worksheet. Cut out the clock face and the two clock hands.

2. Glue the clock face to the center of a paper plate.

3. Line up the dot on the clock face with the dot on each clock hand.

4. Have an adult help you attach the face and two hands together at the dots with a brass paper fastener.

5. Move the hands around the clock to make different times.

6. Write your name on the back of your clock.

7. Take a copy of the Match the Clocks worksheet and write your name on it.

8. Draw a line from each clock on the left to the clock on the right with the matching time.

9. If you have time, select a copy of the What Time? worksheet and complete it by circling the time each clock is showing.

10. Put your clock and completed worksheets in the designated area.

## Follow-Up

• Call up each child and date his or her clock and worksheets.
• Ask questions to determine the child's level of understanding.
• Further explain or demonstrate concepts as needed.
• Record the child's progress.
• Keep the clock and worksheets in the child's folder or send them home.

# Clock Pattern

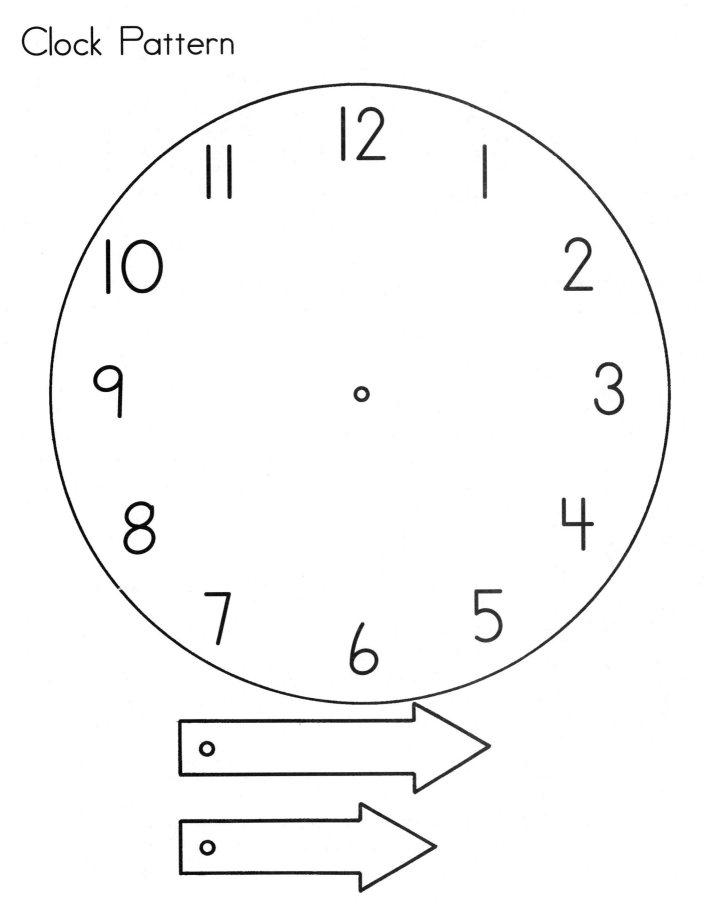

# Match the Clocks

Name_____

Draw a line between the clocks with the same time.

Child understands: ❑ matching.

# What Time?

Name_____

For each clock, circle the time it's showing.

Child understands: ❏ hourly time; ❏ identifying numerals.

# Money Fun

## Objectives

*Practice sorting and identifying coins.*

## Materials Needed

❑ coins
❑ basket
❑ placemat
❑ pencil
❑ worksheets

## Setting Up the Station

• Collect at least five each of pennies, nickels, dimes, and quarters.
• Put the coins in a basket.
• Find a cloth or vinyl placemat and set it on the table in the station.
• Copy the worksheets on pages 135–137.
• Set out the basket of coins, a pencil, and copies of the worksheets.

## The Project

*Explain the following steps to your children.*

1. Empty the coins onto the placemat.

2. Sort the coins by type. Count the number of coins in each pile.

3. Put the coins back into the basket.

4. Take a copy of the Coin Match worksheet and write your name on it.

5. Say the name of the coin at the beginning of the first row. Circle all the matching coins in that row. Repeat for the remaining rows.

6. Select a copy of the Matching Coin Purses worksheet and write your name on it.

7. Look at the coins in the first purse on the left side. Find the purse with the same coins in it on the right side, and draw a line between them. Repeat for the remaining purses.

8. If there is time, take a copy of the How Much? worksheet and write your name on it. Complete the worksheet by counting up the coins in each box and circling the correct amount of money.

9. Place your worksheets in the designated area.

## Follow-Up

• Call up each child and date his or her worksheets.
• Ask questions to determine the child's level of understanding.
• Further explain or demonstrate concepts as needed.
• Record the child's progress.
• Keep the worksheets in the child's folder or send them home.

# Coin Match

Name_____

For each row, name the first coin. Circle the coins that match.

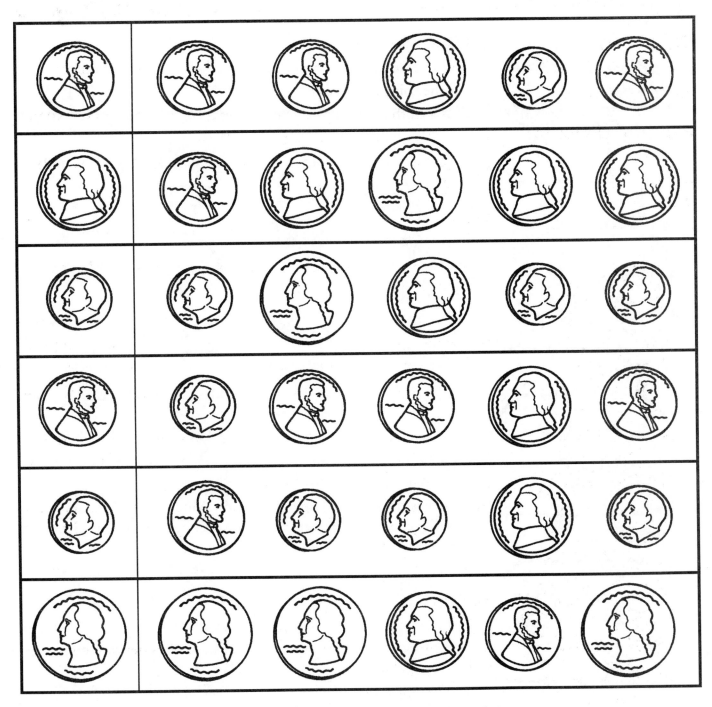

Child understands: ❏ identifying coins; ❏ matching.

# Matching Coin Purses

Name_____

Draw a line between the coin purses with the matching coins.

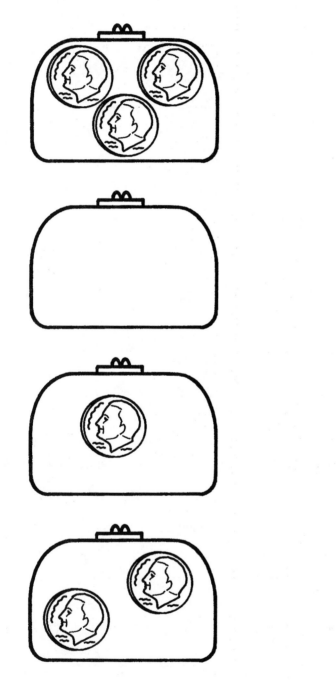
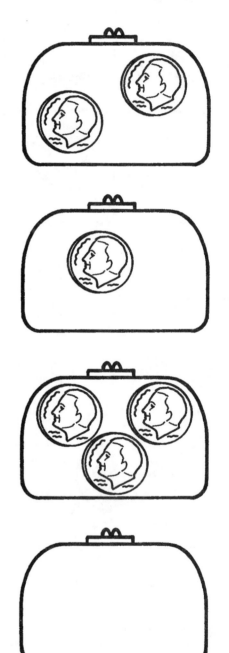

Child understands: ❑ matching.

# How Much?

Name_____

In each box, count up the pennies. Circle the correct amount of money.

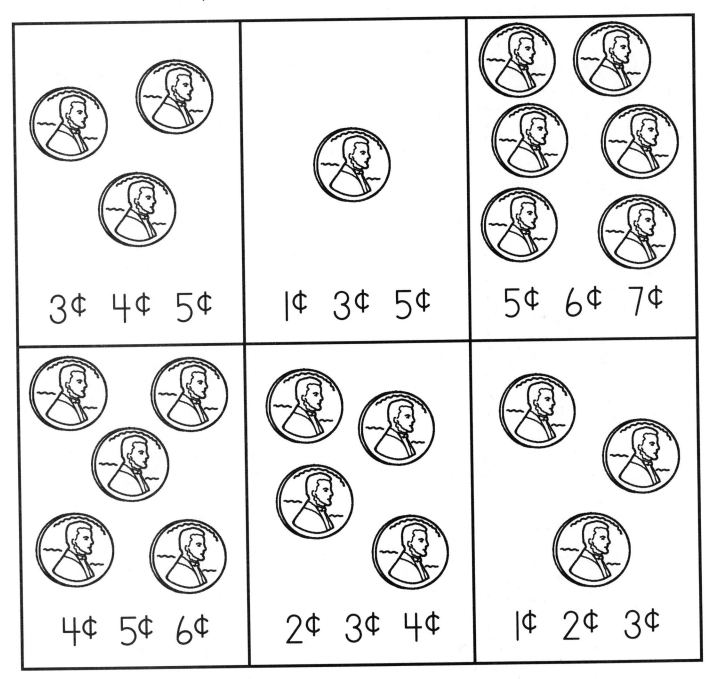

Child understands: ❏ counting pennies; ❏ identifying numerals.

# Half or Whole

## Objectives

*Identify whole and half shapes.*

## Materials Needed

❑ construction paper

❑ scissors

❑ box

❑ pencil

❑ crayons

❑ worksheets

## Setting Up the Station

• Cut pairs of shapes, such as circles, squares, rectangles, triangles, hearts, diamonds, and stars, out of construction paper.

• For each pair, cut one of the shapes in half, keeping both halves.

• Put the whole and half shapes in a box.

• Copy the worksheets on pages 139–141.

• Set out the box of paper shapes, a pencil, crayons, and copies of the worksheets.

## The Project

*Explain the following steps to your children.*

1. Take the paper shapes out of the box.

2. Place the whole shapes together in one pile and the half shapes together in another pile.

3. Put the half shapes together to make whole shapes. Place the matching shapes together.

4. Mix up the paper shapes and put them back in the box.

5. Select a copy of the Half Shapes worksheet, write your name on it, and color half of each shape.

6. Take a copy of the Whole and Half worksheet and write your name on it.

7. Color the whole shapes on the top half of the page and the half shapes on the bottom half of the page.

8. If there is time, select a copy of the Complete the Shapes worksheet, and complete it by drawing in the missing half of each shape.

9. Place your worksheets in the designated area.

## Follow-Up

• Call up each child and date his or her worksheets.

• Ask questions to determine the child's level of understanding.

• Further explain or demonstrate concepts as needed.

• Record the child's progress.

• Keep the worksheets in the child's folder or send them home.

# Half Shapes

Name_____

Color half of each shape.

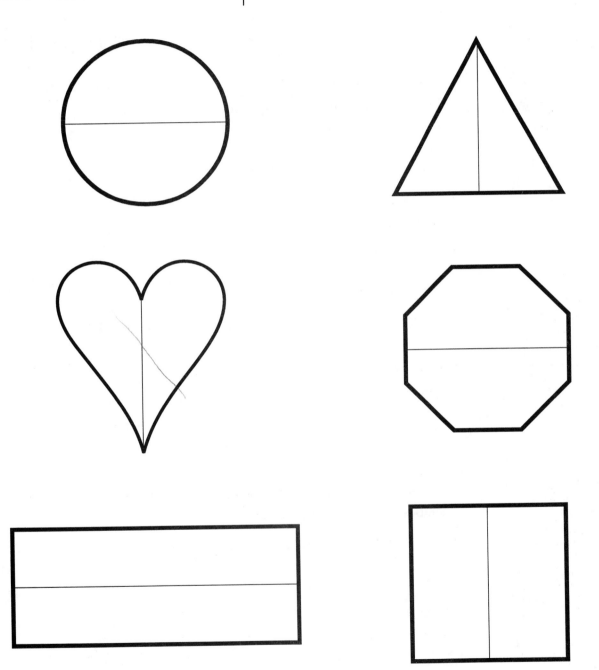

Child understands: ❏ identifying halves.

# Whole and Half

Name_____

Color the shapes that are whole.

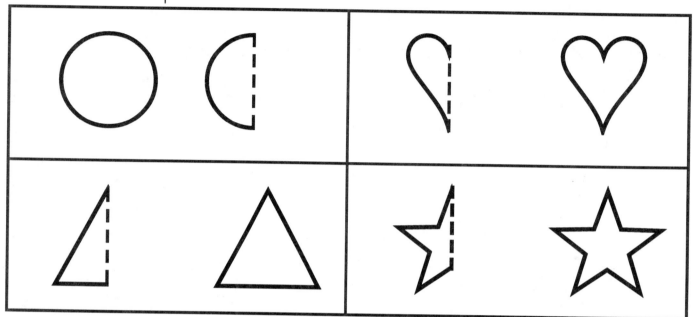

Color the shapes that are half.

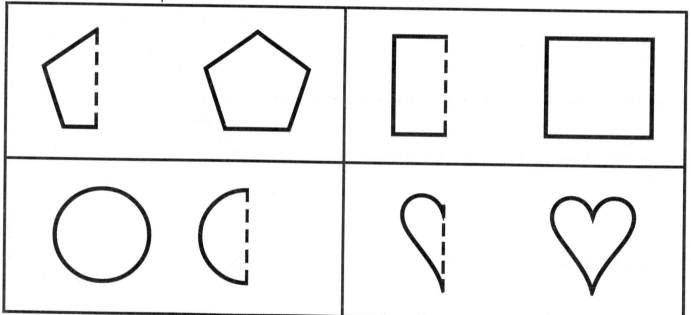

Child understands: ❑ identifying wholes;  ❑ identifying halves.

# Complete the Shapes

Name_____

Complete each shape to make it whole.

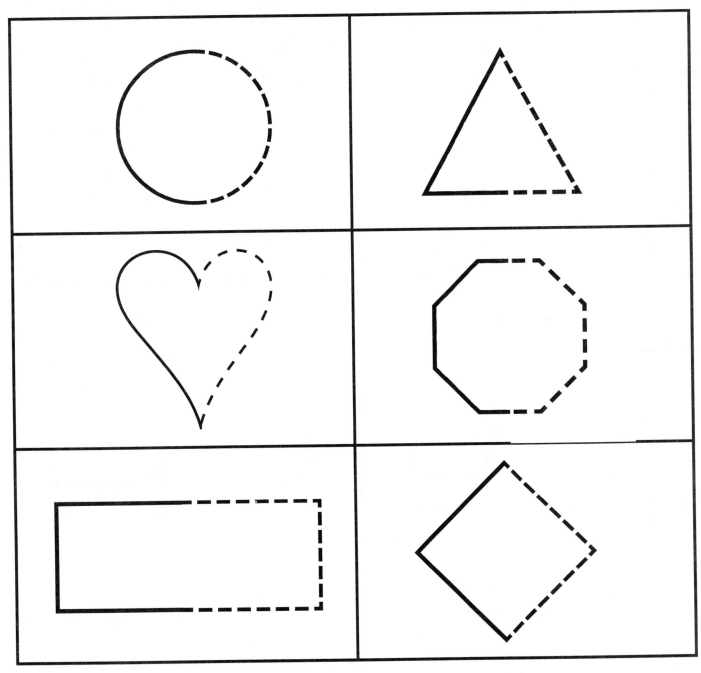

Child understands: ❑ identifying halves; ❑ identifying wholes.

# Stick Puzzles

## Objectives

*Practice completing and making puzzles.*

## Materials Needed

- ❏ worksheets
- ❏ clear self-stick paper
- ❏ craft sticks
- ❏ box
- ❏ pencil
- ❏ construction paper

## Setting Up the Station

- Make one copy of each of the puzzle pattern worksheets on pages 143–145.
- Cover the puzzle worksheets with clear self-stick paper (or have them laminated).
- Purchase small craft sticks (available at craft stores) and put them in a box.
- Set out the puzzle worksheets, the box of craft sticks, a pencil, and construction paper.

## The Project

*Explain the following steps to your children.*

1. Select one of the puzzles. Arrange the craft sticks on the outlines to complete the puzzle.

2. Repeat with the remaining puzzles.

3. Make your own craft stick puzzle. Select a sheet of construction paper and write your name on it.

4. Arrange several of the craft sticks on the construction paper to make a picture. Trace around the craft sticks, then put them back in the box.

5. Try to complete your own puzzle.

6. Place your puzzle in the designated area.

## Follow-Up

- Call up each child and date his or her puzzle.
- Ask questions to determine the child's level of understanding.
- Further explain or demonstrate concepts as needed.
- Keep the puzzle in the child's folder or send it home.

# House Puzzle

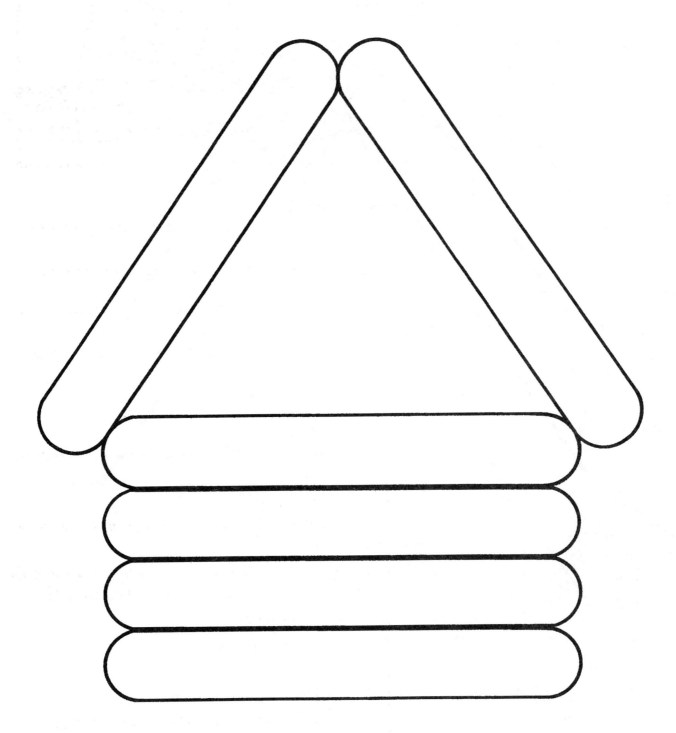

# Tree Puzzle

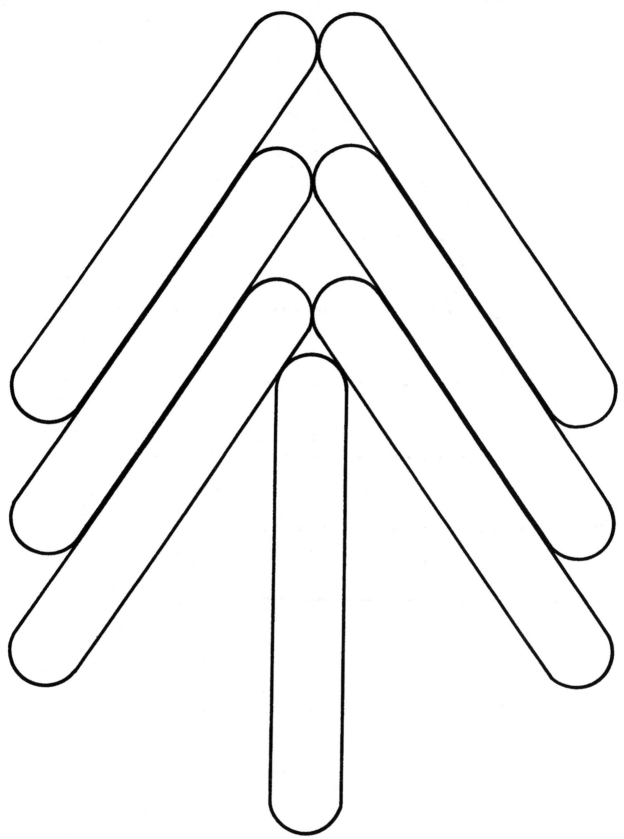

# Stair Puzzle

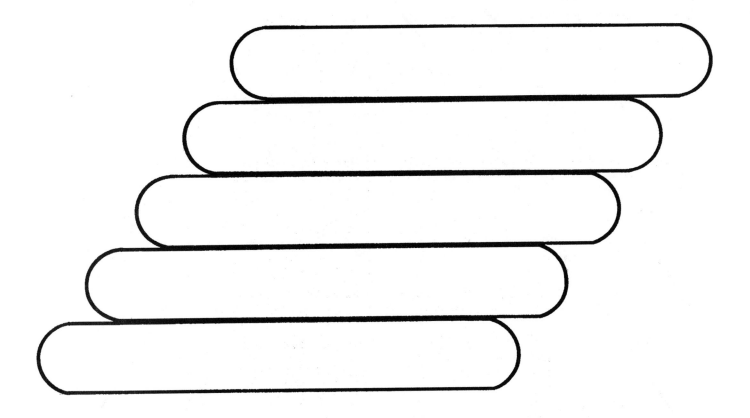

# My Number Book

# My Number Book

*This book gives your children a chance to showcase their knowledge of numbers and to practice important number skills such as identifying numerals, writing numerals, and counting.*

## Directions for Completing the Pages

1. Prepare a folder for each child so you can keep his or her pages together for future binding.

2. Of the following 11 book pages, choose the numeral that you would like your children to work on today. Date the page and make a photocopy for each child.

3. Pass out the copies of the selected page to the children.

4. Explain what to do in each part of the page. At the top of the page, they circle the little numerals that match the big numeral. In the middle of the page, they draw the appropriate number of items pictured. At the bottom of the page, they practice writing the numeral shown.

5. Let the children work on their page.

6. As each child finishes, have him or her bring the page to you. If the work is not complete, encourage the child to finish it.

7. Place each child's page in his or her folder.

## Directions for Making the Book

1. Have the children help you arrange their papers in numerical order.

2. Add a construction paper cover to each set of pages, and staple them together to make a book.

3. Let the children decorate the covers of their books.

4. Send the books home for everyone to enjoy.

Circle all the 1s.

0 1 6 9 | 1 | 6 3 1 4

8 6 5 | | 5 2 7

4 1 7 3 | | 1 4 0 1

Draw 1 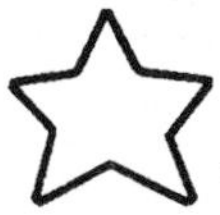.

Write 1.

1 1

Circle all the 2s.

2 0 2 9     6 3 8 4

8 6 5    **2**    5 2 7

4 1 2 3     2 4 0 1

Draw 2  .

Write 2.

2 2

Circle all the 3s.

8 2 3 9     6 3 8 4

3 6 5     **3**     5 2 3

4 1 7 3     3 4 0 1

Draw 3 .

Write 3.

3

Circle all the 4s.

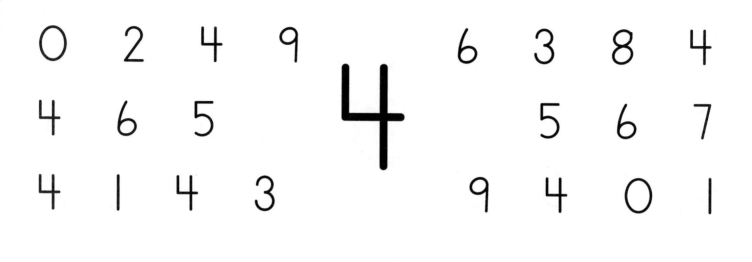

---

Draw 4 ♡.

---

Write 4.

4    4̈

---

Circle all the 5s.

1   2   6   5        6   3   8   5

8   6   5        5   2   7

4   7   5   3        9   4   5   1

5

---

Draw 5 🍎.

---

Write 5.

5   5

Circle all the 6s.

2 7 6 9     6 3 8 4

8 6 5    **6**    5 2 7

4 1 7 3     9 4 6 1

---

Draw 6  .

---

Write 6.

6   6

Circle all the 7s.

5 2 6 9     7     6 3 7 4

8 6 7         5 2 7

7 1 7 3       9 4 0 1

---

Draw 7 .

---

Write 7.

7 7

---

Circle all the 8s.

0   8   6   9       8   3   8   4

8   6   5     **8**     5   2   7

4   1   8   3       9   4   0   8

---

Draw 8   ⬦ .

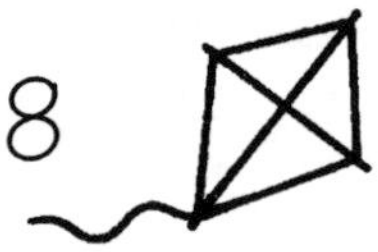

---

Write 8.

8   8

Circle all the 9s.

9 2 6 9     6 3 9 4

              9

8 6 9        5 2 7

9 1 7 3     9 4 0 1

---

Draw 9   .

---

Write 9.

9   9

Circle all the 10s.

| | | | | | | | | |
|---|---|---|---|---|---|---|---|---|
| 0 | 2 | 6 | 9 | | 6 | 3 | 8 | 10 |
| 8 | 10 | 5 | | **10** | 5 | 10 | 7 | |
| 4 | 1 | 7 | 3 | | 9 | 4 | 10 | 1 |

---

Draw 10 .

---

Write 10.

10 10

Circle all the Os.

0  2  6  9       6  3  0  4

8  6  5      O       5  2  7

4  0  7  3       0  4  0  1

---

Draw 0 .

---

Write 0.

# Totline Publications

## Teacher Books

### BEST OF TOTLINE® SERIES
*Totline Magazine's best ideas.*
Best of Totline
Best of Totline Parent Flyers

### BUSY BEES SERIES
*Seasonal ideas for twos and threes.*
Busy Bees—Fall
Busy Bees—Winter
Busy Bees—Spring
Busy Bees—Summer

### CELEBRATIONS SERIES
*Early learning through celebrations.*
Small World Celebrations
Special Day Celebrations
Great Big Holiday Celebrations
Celebrating Likes and Differences

### EXPLORING SERIES
*Versatile, hands-on learning.*
Exploring Sand
Exploring Water
Exploring Wood

### FOUR SEASONS
*Active learning through the year.*
Four Seasons—Art
Four Seasons—Math
Four Seasons—Movement
Four Seasons—Science

### GREAT BIG THEMES SERIES
*Giant units designed around a theme.*
Space • Zoo • Circus

### KINDERSTATION SERIES
*Learning centers for learning with language, art, and math.*
Calculation Station
Communication Station
Creation Station

### LEARNING & CARING ABOUT
*Teach children about their world.*
Our World • Our Town

### MIX & MATCH PATTERNS
*Simple patterns to save time!*
Animal Patterns
Everyday Patterns
Holiday Patterns
Nature Patterns

### 1•2•3 SERIES
*Open-ended learning.*
1•2•3 Art
1•2•3 Blocks
1•2•3 Games
1•2•3 Colors
1•2•3 Puppets

1•2•3 Reading & Writing
1•2•3 Rhymes, Stories & Songs
1•2•3 Math
1•2•3 Science
1•2•3 Shapes

### 101 TIPS FOR DIRECTORS
*Valuable tips for busy directors.*
Staff and Parent Self-Esteem
Parent Communication
Health and Safety
Marketing Your Center
Resources for You
   and Your Center
Child Development Training

### 101 TIPS FOR PRESCHOOL TEACHERS
Creating Theme
   Environments
Encouraging Creativity
Developing Motor Skills
Developing Language Skills
Teaching Basic Concepts
Spicing Up Learning Centers

### 101 TIPS FOR TODDLER TEACHERS
Classroom Management
Discovery Play
Dramatic Play
Large Motor Play
Small Motor Play
Word Play

### 1001 SERIES
*Super reference books.*
1001 Teaching Props
1001 Teaching Tips
1001 Rhymes & Fingerplays

### PIGGYBACK® SONG BOOKS
*New lyrics sung to the tunes of childhood favorites!*
Piggyback Songs
More Piggyback Songs
Piggyback Songs for Infants
   and Toddlers
Holiday Piggyback Songs
Animal Piggyback Songs
Piggyback Songs for School
Piggyback Songs to Sign
Spanish Piggyback Songs
More Piggyback Songs for School

### PROBLEM SOLVING SAFARI
*Teaching problem solving skills.*
Problem Solving—Art
Problem Solving—Blocks
Problem Solving—Dramatic Play
Problem Solving—Manipulatives
Problem Solving—Outdoors
Problem Solving—Science

### SNACKS SERIES
*Nutrition combines with learning.*
Super Snacks • Healthy Snacks
Teaching Snacks • Multicultural Snacks

### THEME-A-SAURUS® SERIES
*Classroom-tested, instant themes.*
Theme-A-Saurus
Theme-A-Saurus II
Toddler Theme-A-Saurus
Alphabet Theme-A-Saurus
Nursery Rhyme Theme-A-Saurus
Storytime Theme-A-Saurus
Multisensory Theme-A-Saurus

### TODDLER SERIES
*Great for working with 18 mos–3 yrs.*
Playtime Props for Toddlers
Toddler Art

## Tot-Mobiles
*Unique sets of die-cut mobiles for punching out and easy assembly.*
Animals & Toys
Beginning Concepts
Four Seasons

## Puzzles & Posters

### PUZZLES
Kids Celebrate the Alphabet
Kids Celebrate Numbers
African Adventure
Underwater Adventure
Bear Hugs 4-in-1 Puzzle Set
Busy Bees 4-in-1 Puzzle Set

### POSTERS
We Work and Play Together
Bear Hugs Health Posters
Busy Bees Area Posters
Reminder Posters

## Story Time
*Delightful stories with related activity ideas, snacks, and songs.*

### KIDS CELEBRATE SERIES
Kids Celebrate the Alphabet
Kids Celebrate Numbers

## Parent Books

### A YEAR OF FUN SERIES
*Age-specific books for parenting.*
Just for Babies
Just for Ones
Just for Twos
Just for Threes

Just for Fours
Just for Fives

### BEGINNING FUN WITH ART
*Introduce your child to art fun.*
Craft Sticks • Crayons • Felt
Glue • Paint • Paper Shapes
Modeling Dough • Tissue Paper
Scissors • Rubber Stamps
Stickers • Yarn

### BEGINNING FUN WITH SCIENCE
*Spark your child's interest in science.*
Bugs & Butterflies • Plants & Flowers
Magnets • Rainbows & Colors
Sand & Shells • Water & Bubbles

### LEARN WITH PIGGYBACK® SONGS BOOKS AND TAPES
*Captivating music with age-appropriate themes help children learn.*
Songs & Games for Babies
Songs & Games for Toddlers
Songs & Games for Threes
Songs & Games for Fours
Sing a Song of Letters
Sing a Song of Animals
Sing a Song of Colors
Sing a Song of Holidays
Sing a Song of Me
Sing a Song of Nature
Sing a Song of Numbers

### LEARN WITH STICKERS
*Beginning workbook and first reader with 100-plus stickers.*
Balloons • Birds • Bows • Bugs
Butterflies • Buttons • Eggs • Flags
Flowers • Hearts • Leaves • Mittens

### LEARNING EVERYWHERE
*Discover teaching opportunities everywhere you go.*
Teaching House
Teaching Trips
Teaching Town

### SEEDS FOR SUCCESS
*Ideas to help children develop essential life skills for future success.*
Growing Creative Kids
Growing Happy Kids
Growing Responsible Kids
Growing Thinking Kids

### TIME TO LEARN
*Ideas for hands-on learning.*
Colors • Letters • Measuring
Numbers • Science • Shapes
Matching and Sorting • New Words
Cutting and Pasting
Drawing and Writing • Listening
Taking Care of Myself